INFORMED CONSENT, PROXY CONSENT, AND CATHOLIC BIOETHICS

Philosophy and Medicine

VOLUME 112

CATHOLIC STUDIES IN BIOETHICS

INFORMED CONSENT, PROXY CONSENT, AND CATHOLIC BIOETHICS

For the Good of the Subject

by

GRZEGORZ MAZUR, O.P.

Dominican College of Philosophy and Theology, Kraków, Poland

Springer

Grzegorz Mazur, O.P.
Dominican College of Philosophy
and Theology
Stolarska 12
31-043 Kraków
Poland
grzegorz.mazur@dominikanie.pl

ISSN 0376-7418
ISBN 978-94-007-2195-1 e-ISBN 978-94-007-2196-8
DOI 10.1007/978-94-007-2196-8
Springer Dordrecht Heidelberg London New York

Library of Congress Control Number: 2011935463

Printed on acid-free paper

Springer is part of Springer Science+Business Media (www.springer.com)

Foreword

It is a pleasure and a privilege to write this Foreword. Father Mazur's study not only offers a study of value for "Catholic" bioethics but also one of great significance for contemporary bioethics as such. The problem that he explores is one with which I have wrestled since 1976 when I wrote "Proxy Consent to Human Experimentation" for *The Linacre Quarterly* stimulated several responses, and I developed the argument advanced in my 1977 book *Human Existence, Medicine, and Ethics.* I was compelled much later, because of what I came to realize were valid objections to my position, to change it significantly. Because of my long interest in this issue I thought that I had a good grasp of relevant literature. The truly amazing research that Father Mazur carried out in writing this superlative study showed me how ignorant I was of very important contributions to this literature.

Mazur had in fact set forth accurately and brilliantly criticized all of the most important positions held by bioethicists on this matter. His examination, for example, of the widely used, highly influential, and very complex thought of Tom Beauchamp and James Childress in their *Principles of Biomedical Ethics*, is one of the finest I have ever seen. He offers a similarly superb examination of the famous debate during the late 1970s and early 1980s between Paul Ramsey and Richard McCormick, S.J. on the grounds for giving proxy consent on behalf of incompetent persons—those Ramsey called "voiceless subjects." The same is true of his critical analysis of the pure autonomy view that undermines human dignity and his rejection of the "minimal risk" standard unfortunately widely accepted in our culture, the "best interest of the patient" standard, and similar views.

Mazur roots his study in a sound anthropological and moral theory, one fully compatible with the authoritative teaching of the Church and supported by sound philosophical analysis. Note that I mention both "anthropological" and "moral" theory. Although moral theory is not, as some hold, *derived* from anthropology or one's understanding of a human person, anthropology and moral theory are inherently interrelated, as Mazur shows. Indeed, one of the great truths magnificently developed in this important study is that a sound morality goes hand in hand with a sound anthropology; although moral theory is *epistemologically* or *gnoseologically* independent of anthropology (one cannot derive an "ought" from an "is"), moral theory is *ontologically* rooted in anthropology.

In short, this is an exceptionally valuable study; its contribution to bioethics is outstanding.

William E. May, Ph.D.

Introduction

The principle of free and informed consent enjoys broad recognition in present-day bioethics as a cardinal principle governing scientific research with human subjects. The principle gained worldwide prominence in the wake of World War II, as a result of the appalling experiments carried out by Nazi physicians on concentration camp prisoners and prisoners of war, under the pretence of scientific progress. As this study will show, however, Catholic thought acknowledged the significance of free and informed consent long before World War II. This acknowledgment traces back to the pioneering contributions of the Catholic doctors associated with the nineteenth-century French school of *Morale Médicale*.

Despite widespread acknowledgment, free and informed consent is nevertheless a conceptually and morally problematic principle. What exactly do the terms "free" and "informed" connote? Is informed consent the only principle relevant in the topic of research with human subjects? In what precisely does it consist? Can it be viewed as a moral absolute? How must it be understood in order to safeguard the dignity of human subjects participating in research? And, perhaps most importantly, is the self-determination the principle makes possible that of a self-legislating autonomy, or a freedom guided by an objective good?

Similar questions, and more, plague discussion and application of the principle of proxy, surrogate, or vicarious consent. This kind of consent is engaged when the competence of patients or research subjects is compromised or nonexistent. It is widely agreed that in such situations proxies are entitled to offer binding consent to experimentation on their wards. But on what grounds is doing so justified, and what are the boundaries of such consent? Furthermore, how does proxy consent in therapeutic research with human subjects differ from such consent in nontherapeutic settings? Is this difference relevant from the point of view of sound anthropology and moral justification?

As this study will show, various reasons have been proposed to explicate the meaning of the principle of free and informed consent and the justification of proxy consent for experimenting on human subjects. In some areas the answers are still being sought. It is far from evident, for instance, on what premises the justification of proxy consent to therapeutic research rests. Likewise, sharp disputes exist—even among Catholic moral theologians—about whether proxy consent can ever be deemed morally valid in nontherapeutic settings and, if it can, under what conditions and why.

One of the most hotly debated issues regards the notions of *minimal risk* and *slight increase over minimal risk* in nontherapeutic research. Put briefly, is it licit to subject the incompetent person to nontherapeutic procedures involving minimal risk or a risk slightly exceeding minimal? If so, what benchmark should be employed to ascertain that the risk at stake is minimal or only slightly higher than minimal? If not, how might it be plausibly argued that any amount of risk in nontherapeutic experimentation renders it morally objectionable? Would this argument affect therapeutic research as well? What degree of risk can be justified in therapeutic settings and how does this justification relate to the risk-benefit ratio to be determined in each therapeutic research project?

While answering these and other questions, this study will offer analysis and critique of the major positions advanced to justify informed and proxy consent to both therapeutic and nontherapeutic experimentation; a sounder understanding of the foundations of both informed and proxy consent will be presented through the incorporation of Catholic anthropological and moral premises.

The first chapter will cover the historical origins of informed consent while the second reviews its articulation in human-rights documents. These will provide an indispensible background for the following chapter, in which the issue of informed consent will be traced throughout the Magisterial teaching of the Catholic Church, as well as in the relevant philosophical and theological literature. In addition, the topic of the exceptions to informed consent will be surveyed, followed by an introduction to the issue of proxy consent. This introduction will address a number of ambiguities in terminology permeating today's bioethics literature and will defend the distinction between therapeutic and nontherapeutic research, which has been assailed by some authors. With this foundation completed, the stage will be set for the examination of proxy consent in succeeding chapters.

This examination begins with a discussion of the issue of proxy consent in therapeutic research. After the presentation of mainstream accounts of proxy decision-making predominant in the present-day bioethics literature, a comprehensive critique of the principle of autonomy which undergirds these accounts will be offered, emphasizing the anthropological and moral consequences of reducing human freedom to pure self-determination divorced from the objective good. Subsequently, building upon the Magisterial teaching as well as on the philosophical and theological approaches dealt with earlier, a reliable rationale for proxy consent in therapeutic experimentation, harmonious with Christian anthropology and ethics, will be developed.

We will then turn to the debate on the justification of proxy consent in nontherapeutic research and will critically assess its main rationales. As a basis for this scrutiny, the examination of this topic will first consider different types of nontherapeutic research viewed through the lens of basic research taxonomy. In order to foster a more comprehensive understanding of the intricacies of nontherapeutic research, it will also survey the four categories of research subjects who are especially vulnerable in such research. In addition, the issues of embryonic stem cell research, genetic therapy, and prenatal and pre-implantation screening will be considered under the category of unborn research subjects. Finally, the examination of the relevant Magisterial teaching along with two theological approaches in harmony with it will complete chapter four.

With this groundwork in place for the resolution of the issue of the moral legitimacy of proxy consent, the study will continue with a summary aimed at a comprehensive understanding of proxy consent in light of the principles of objective morality such as truth, conscience, and virtue. While not neglecting the therapeutic setting, this review will focus on the risk factor critical in the moral evaluation of nontherapeutic research. Finally, the qualified justification of proxy consent in such research will be developed by pointing to the relationship between the individual good of incompetent subjects and the common good in which they share.

Contents

Abbreviations

AAS	Acta Apostolicae Sedis
Acad Emerg Med	Academic Emergency Medicine
Am J Public Health	American Journal of Public Health
Ann Intern Med	Annals of Internal Medicine
Behav Sci Law	Behavioral Science & the Law
BMJ	British Medical Journal
Can J Psychiatry	The Canadian Journal of Psychiatry
CHAC	Catholic Health Association of Canada
Crit Care Med	Critical Care Medicine
JAMA	The Journal of the American Medical Association
J Med Ethics	Journal of Medical Ethics
NCBQ	The National Catholic Bioethics Quarterly
New England J Med	The New England Journal of Medicine
Obstet Gynecol	Obstetrics & Gynecology
SAMJ	South African Medical Journal

Chapter 1
The Historical Development of the Principle of Free and Informed Consent

In medical ethics there are two basic contexts in which one commonly speaks of informed consent: the first is the context of treatment, while the second is that of research. Informed consent in both these contexts demonstrates no essential dissimilarities, yet it differs in regard to its "original recognition, evolution and development" as well as its "means of enforcement" (Nelson-Marten and Rich 1999, 82). Indeed, "despite apparent similarities . . . informed consent in the research setting has evolved quite separately from informed consent to treatment." The latter "is largely a creature of case law, with some subsequent statutory modifications" whereas the former "has been shaped by professional codes, statutes, and administrative regulations with the courts playing a less important role" (Berg et al. 2001, 249; see also Beauchamp and Faden 2004, 1274). Although by no means as critical for this work as its research counterpart, informed consent to treatment will be invoked occasionally for either comparative or supplementary purposes. This seems necessary for better understanding of the subtleties of consent and the actual weight of issues related to it.

1.1 Debate on the Origin of the Principle of Free and Informed Consent in Medical and Research Practices

The origin of the concept of consent in both treatment and research settings remains subject to animated debate. The point of contention consists primarily in the methodology employed in tracing the origins of what we call informed consent. One group of scholars begins with a current understanding of consent and examines its existence throughout history. Another group attempts to locate different concepts and policies on decision-making in use over the centuries and to scrutinize them in light of the standards of the time.

The principal representative of the former approach, psychiatrist Jay Katz, in his monumental work *Experimentation with Human Beings*[1] offers for consideration the 1767 case *Slater v. Baker and Stapleton, C.B.*, in which an English court declared "improper" a procedure performed in relation to the regular treatment of a broken leg. The treatment was experimental and clearly contrary to the will of the patient.

G. Mazur, O.P., *Informed Consent, Proxy Consent, and Catholic Bioethics*, Philosophy and Medicine 112, DOI 10.1007/978-94-007-2196-8_1,

Moreover, it was "ignorance and unskilfulness in that very particular, to do contrary to the rule of the profession, what no surgeon ought to have done; and indeed it is reasonable that a patient should be told what is about to be done to him." The court also accepted the witness of a physician underscoring the importance of the patient's consent (Katz 1972, 527).

This evidence, however, seems insufficient to establish that a fully-developed notion of consent existed at the end of the eighteenth century. In fact, Katz couples the genesis of informed consent only with the 1960 *Natanson v. Kline* case, in which a physician was sued and found guilty for not securing informed consent for cobalt radiation therapy. Although the patient consented, he was not informed. In other words, regardless of having obtained consent, the practitioner did not fully notify the patient about the nature of the procedure and associated risks. Interestingly, the disclosure requirement brought up by the court in *Natanson* can be found—although in less explicit manner—as early as in the aforementioned *Slater.*

Katz's approach can best be characterized according to his own words, which express general disbelief in both the definition and the application of the concept of informed consent. Katz writes, "law has neither defined sufficiently well the substance and ambit of informed consent in therapeutic settings nor determined clearly its functional relevance for human experimentation. Thus, in invoking informed consent like a talisman, lawyers, investigators, and courts often seem to overlook the fact that it lacks specific construction and remains an ill-defined concept" (Katz 1972, 523). In light of these words one can legitimately question whether Katz acknowledges that informed consent as such even exists at this point. Even if it is true that the term was invented and used in legal settings, it does not appear to be well enough connected with and incorporated into reality. In other words, what Katz actually believes can be put as follows: it is only a premature and disorganized consent rather than its genuine form that thus far has been brought into existence.

Katz's skepticism finds further support in his later book, *The Silent World of Doctor and Patient* in which the origin of informed consent has been associated with the legal case that coined and for the first time used the term "informed consent." As Katz put it, "the doctrine of informed consent surfaced, seemingly out of nowhere, in *Salgo v. Leland Stanford Jr. University Board of Trustees*" (Katz 2002, 60; quoted in Faden, Beauchamp, and King 1986, 58) in 1957. The phrase "out of nowhere" advances the claim that before 1957 there was simply no consent, at least in its "informed" shape. Yet even after *Salgo* "the stark silence that throughout medical history has been the hallmark of physicians' attitude toward patients' participation in medical decision making" did not disappear. Instead, "the physician-patient dialogue, now a mixture of proffered and withheld information, has become even more opaque; surely it is not based on the idea of shared decision making" (Katz 2002, 3). In other words, even if the noble idea of informed consent had been introduced by the courts, it did not alter the physicians' mindset and can hardly be regarded as being backed up by their actions.

Indeed, one of the purposes of Katz's project is to document that "disclosure and consent, except in the most rudimentary fashion, are obligations alien to medical thinking and practice" (Katz 2002, 1). In fact, "the history of the physician-patient relationship from ancient times to the present ... bears testimony to physicians' inattention to the right and need of their patients to make their own decisions. Little appreciation of disclosure and consent can be discerned in this history, except negatively, in the emphasis on the inability of patients to grasp the mysteries of medicine and therefore to share the burdens of decision with their doctors" (Katz 2002, 28).

While Katz's views lie at one pole in the debate, those of historian Martin Pernick occupy a recognized place at the other. Pernick's research indicates that "truth-telling and consent-seeking have long been part of an indigenous medical tradition, based on medical theories that taught that knowledge and autonomy had demonstrably beneficial effects on most patients' health" (Pernick 1982, 3; quoted in Faden, Beauchamp, and King 1986, 56). Despite the different social context of the nineteenth century, "Pernick believes it is unassailable that meaningful consent practices existed in American medicine" as early as that period (Faden, Beauchamp, and King 1986, 56). In support of this claim, the Episcopalian theologian Joseph Fletcher maintains that "the moral obligation of informed consent in medicine and research was reasonably clear before 1939," and that the courts had already been active before that time in "establishing the validity of the principle" (Fletcher 1983, 204; quoted in Faden, Beauchamp, and King 1986, 56). Pernick's and Fletcher's position appears buttressed "by the well-entrenched belief in legal writings that the law of informed consent can be traced at least to the early days of the twentieth century" (Faden, Beauchamp, and King 1986, 56).

In the face of these two approaches, apparently in opposition to each other, an attempt has been made to introduce a new, more balanced theory regarding the beginnings of informed consent. In their book *A History and Theory of Informed Consent*, Ruth Faden and Tom Beauchamp[2] propose to reconcile Katz's thesis with that of Pernick. After admitting that "where Katz sees no informed consent, Pernick finds it in abundance," (Faden, Beauchamp, and King 1986, 57) they suggest that a plausible reason why the two substantially disagree on the matter is the aforementioned diverse methodology utilized in their undertakings. This has also been conceded by Pernick himself who charged Katz with "tracing the history of 'informed consent' as if it were an isolated entity that could be abstracted from its historical context."[3] Taking that context into account, Katz's thesis on the "history of silence" does not necessarily rule out Pernick's full confidence that both disclosure and consent were established in medical practice in earlier times.

Nevertheless, even if the latter is true and the two opponents are not quite poles apart, it still must be acknowledged that real differences exist between their positions. According to Katz, for instance, the sources Pernick cites provide proof of bringing patients to consent rather than enabling them to give their own informed consent. To put it another way, by disclosure of information practitioners endeavored to make patient consent congruent with what they thought should be done.

Being aware of the clearly dissimilar conclusions reached by Pernick and Katz, Faden and Beauchamp propose to define at the outset what they intend to focus on in the history of medicine. The criteria for informed consent should be neither too constrained nor too loose, lest the risk of inaccurate results jeopardize the entire project. Aware of this danger, our authors believe that the following three conditions make a fair and realistic yardstick for informed consent when exploring historical data. First, "a patient or subject must agree to an intervention based on an understanding of (usually disclosed) relevant information." Second, "consent must not be controlled by influences that would engineer the outcome." Third, "the consent must involve the intentional giving of permission for an intervention."

In addition, one relevant distinction needs to be introduced between what has been labeled *a failure to satisfy criteria* and *a failure of evidence*. The mere absence of evidence supporting the practice of informed consent does not suffice to verify its nonexistence in a particular place or time. Faden and Beauchamp admit that "it is sometimes tempting to hold that there was no practice of informed consent during a given period because our criteria are not satisfied, when all we can say in fairness to the facts is that we do not have good and sufficient evidence for saying that at the time informed consent either did or did not exist" (Faden, Beauchamp, and King 1986, 54). Moreover, the available historical evidence is not always of the highest quality. Likewise, the investigator rarely knows to what extent, if any, the records found represent typical, large-scale, and recognized practices of the day. Some pieces of data may refer to isolated events rather than to routine and recognized medical procedures.

In contrast to their moderate and balanced introductory distinctions and despite their occasional disagreement with Katz, Faden and Beauchamp adopt his methodology, applying the current notion of consent to former times and expressing a strong criticism regarding the approach based on the history of informed consent. In the paragraph opening the section on the history of the concept in clinical medicine, they state that this history "shows primarily how inadequately, and with what measure of hostility and insularity, problems of truthfulness, disclosure, and consent were framed and discussed prior to the twentieth century" (Faden, Beauchamp, and King 1986, 60). By analogy, in the domain of experimentation, they note that although "the practice of research with human subjects is virtually as ancient as medicine itself . . ., concern about its consequences and about the protection of human subjects is a recent phenomenon" (Faden, Beauchamp, and King 1986, 151). This statement, despite its radical character, goes unsupported by any reference, evidence, or explanatory note. Thus, it seems to have been simply assumed. Consequently, it remains unclear how it is possible to determine that throughout the centuries, since ancient times, the practice of research lacked concern about subjects and failed to protect them.

Fortunately, another claim proposed by Faden and Beauchamp helps to clarify this issue. They admit somewhat tersely that "events in research ethics before the end of World War II that merit attention are few in number and without significant impact on later developments." Even supposing that this is true, however, it remains unclear exactly how our authors know this. As has been shown, a likely explanation for this phenomenon is the attempt, made by others as well as by Faden and Beauchamp, to associate the origins of genuine informed consent exclusively with the North American legacy—a practice obviously at odds with historical data (Herranz 2004, 133).

The paucity of evidence in Faden and Beauchamp's work does not justify their conclusions and suggestions. This in turn stands in opposition to their prior distinction between *failure to satisfy criteria* and *failure of evidence*; in fact, they draw conclusions as if they possessed the evidence to support them. Yet, in most instances, they honestly admit that only "little evidence" is available with respect to consent in a given period. Therefore, as mentioned above, one can conclude with certitude neither the consent's existence nor its non-existence. This becomes even more obvious when still-occurring discoveries of historic data regarding the matter are taken into account. One such discovery, for instance, refers to the existence in the seventeenth-century Ottoman Empire of "the established practice of legal contracts between physician and patient, which were drawn up and signed in the presence of witnesses" (Ajlouni 1995, 980). Similar contracts can be found in Italy and France as early as in the fourteenth century (Shatzmiller 1994, 129–30). Although not dealing directly with the disclosure issues, those documents draw attention to the fact that consent-based legal agreements between physician and patient have been a part of medical practice at least since the Middle Ages. For that reason, it seems judicious to admit the probability that future findings might show the existence of fully evidenced informed consent. Nonetheless, Faden and Beauchamp give priority to the theory that informed consent is only a twentieth-century product, paralleling Katz's conclusions in this regard.

The problem of the adequacy of conclusions in relation to corresponding historic data, however, is not the only issue that characterizes Faden and Beauchamp's work. A careful reading reveals a number of others that, given the limited scope of this study, cannot be presented in detail. It is not my task to engage in extensive and historically oriented research; nevertheless, I intend to comment briefly on these issues. I believe that Faden and Beauchamp's presentation, in spite of its undeniable value and its position in the professional literature, can be portrayed as an account of various assumptions, usually unsupported by significant historic evidence. Now, I will take a look at three of their assumptions.

The first assumption points to the method adopted by the authors. In summary, they believe that the history of medicine can be legitimately re-described through the prism of two ethical principles, namely that of *respect for autonomy* and that of *beneficence*. In addition to the questionable practice of projecting philosophical premises onto historic data, the difficulty here consists in two elements: the philosophical meaning associated with the two principles, and the tactic of detaching one principle from the other (or even, in some instances, of juxtaposing them). While I will take a closer look at the first component later in this work, I will now share a few thoughts on the second.

Faden and Beauchamp clearly see the two principles as justification for two corresponding models of patient-doctor relationship. One model, derived from the Hippocratic Oath, has been founded on beneficence, whereas the other, allegedly more evident in modern times, has been grounded in autonomy. In other words, the principle of beneficence is fundamental to the former model and the principle of autonomy applies to the latter. In the first model it is chiefly the physician who decides what should be done with respect to the patient's therapy. His goal is to provide benefit to the patient, and the responsibility rests on him. Even if dialogue is used, it does not serve to obtain fully autonomous consent but rather to make the patient comply with what the doctor thinks is best. By contrast, the autonomy model perceives consent as an independent decision on the part of the patient based on the information provided by the doctor. Here the patient is in charge by reason of the principle of autonomy according to which "persons should be free to choose and act without controlling constraints imposed by others" (Faden, Beauchamp, and King 1986, 8). To put it another way, freedom of both choice and act determines autonomy and vice versa. I believe that this relationship between freedom and autonomy is erroneous and I will come back to it in later chapters.

A second assumption made by Faden and Beauchamp is that the two models of medicine can be seen as existing in apparent opposition to each other. Nevertheless, the opposition between the beneficence and the autonomy model with respect to the historic origins of informed consent has been neither explained nor justified. It has simply been assumed. Thus, for instance, no proof has been offered for the claim that the Hippocratic model operational in medical practice for centuries precluded respect for autonomy. This claim is possible only because of the premise that the two models oppose each other. According to the logic of this supposition, if there are more indicators of beneficence in a certain period or practice, Faden and Beauchamp assume that respect for autonomy was diminished or even destroyed. In medical reality, however, as Pernick showed and which also did not escape our authors' notice, physicians "in the eighteenth and nineteenth centuries" respected autonomy "because of the beneficial therapeutic outcome of such research." From this Faden and Beauchamp conclude merely that Pernick agrees with them that "physicians operated out of a beneficence model in the consent context" (Faden, Beauchamp, and King 1986, 60). While this is true, there is also at least one additional conclusion not brought up by our authors. Pernick's research points to the fact that in real medical practice beneficence and respect for autonomy go hand in hand and should be seen as one integrated entity. This, however, stands clearly at odds with their approach, according to which prima facie principles of beneficence and respect for autonomy conflict with each other. Faden and Beauchamp are right that, given their assumptions, under certain conditions it is beneficence that prevails, whereas under others autonomy takes the lead. Yet it is certainly too much to say that those two principles oppose each other. In fact, depending on the moral character of both practitioner and patient, they might be used in such a way as to complement each other.

To put it another way, it is not solely the principle of respect for autonomy that protects the person's dignity and freedom of choice. In the Hippocratic view, the physician was not an adversary against whom the patient needed protection. Accordingly, no threat to autonomy had to be envisioned. This, in my judgment, is precisely what Faden and Beauchamp fail to grasp in their historical account. A possible explanation for this failure is that they relate their theory closely to the legal system. Indeed, it is true that the notion of informed consent originated in courts of law. It also remains true that the right to self-determination, being the legal equivalent of the principle of autonomy in ethics, has played a decisive role in the court-based justification of the doctrine of informed consent. Nevertheless, while working sufficiently well in the legal context, this principle, as will be demonstrated in a more detailed fashion in later chapters, does not satisfactorily articulate the rich complexity of both moral behavior and human dignity.

The third assumption to be found in the work of Faden and Beauchamp concerns different religious traditions and the influence they exert on the history of informed consent. At one point this influence is labeled as "indirect" and seldom "paramount." According to our authors, "while religiously rooted notions such as that of the Christian gentlemen influenced various epochs in medical ethics, it is doubtful that the moral positions that developed internally within the profession of medicine received a theological justification or grew from distinctly religious traditions" (Faden, Beauchamp, and King 1986, 60). However, only a few pages later, in a section devoted to medieval medicine, the authors declare that "Christian monastic physicians traditionally held to the Hippocratic traditions, where medical authoritarianism and obligations of obedience by patients were strengthened further in theology" (Faden, Beauchamp, and King 1986, 63). In other words, the moral position of medical authoritarianism received an additional justification in theology. The latter claim not only contradicts the former but also casts suspicion on what has been called "theology" and on its place in medical practice, in particular by the surprising connection implied between theology and the obligation of obedience. Alas, Faden and Beauchamp neither specify what type of theology they refer to, nor present any source supporting their claim.

An example offered to illustrate the claim that religion had little influence on the history of informed consent shows that Faden and Beauchamp are speaking of theology only very loosely. Their example involves Thomas Percival, an eminent British physician whose 1803 *Medical Ethics* served as a key source for the first American Medical Association Code of Ethics adopted in 1847. Our authors note that Percival's "medical ethics reflects the Christian gentleman conception, yet very little in his pragmatic volume is directly traced or traceable to doctrines of *Christian* ethics. To be a Christian gentleman really meant for Percival that a doctor should be virtuous" (Faden, Beauchamp, and King 1986, 60). Given the fundamental place virtues occupy in Christian ethics as well as the well-documented research done on Percival's writings, this surprising conclusion is far too facile. As Edmund Pellegrino phrased it, in reality "Percival's virtues were based in Christian theology, which supplements the natural virtues of benevolence and beneficence with the supernatural virtue of charity. That conjunction of Greek teaching about the virtues with Christian theology is a recurrent phenomenon since the early Greek and Latin fathers of the Church" (Pellegrino 1985, 22–23). To put it another way, even though some of the virtues Percival held in deep esteem "were not far from the Stoic virtues . . ., the chief characteristic of Percival's life and thought was the close integration of faith and reason. Philosophy and theology were, for him, complementary and not antagonistic as they were to many Enlightenment thinkers." In fact, as Percival attained mature age, "theology came to play the dominant role" in his life (Percival 1985, 19). This was reflected for instance in his view on suffering, illness, and the role of Divine Providence in healing (Percival 1985, 20). Indeed, the link between Percival's ethics and Christian theology is strong and certain.[4] What is more, this link demonstrates Percival's wholehearted dedication to the Christianized form of the Hippocratic ethos in medicine.

The Hippocratic Oath was incorporated into Christianity in the Middle Ages.[5] Consequently, a Christian model of the physician, anchored in theology, has remained highly influential even up to the present time. As a brief historical summary preceding the text of AMA Code of Medical Ethics in its latest edition points out, "the Oath of Hippocrates ... has come down through history as a living statement of ideals to be cherished by the physician. This Oath ... (Christianized in the tenth or eleventh century A.D. to eliminate reference to pagan gods) has remained in Western Civilization as an expression of *ideal conduct for the physician*" (Council on Ethical and Judicial Affairs 2006, xi, emphasis added). In the following section, I will show how Christian theology, operational through the Hippocratic ethos, contributed to the nineteenth-century informed consent claim made by French physicians of the *Morale Médicale* school.

In summary, the above-mentioned difficulties with Faden and Beauchamp's approach call into question the accuracy of the historic part of their book. Furthermore, as Herranz has noted, the practice, embraced by some authors of a similar intellectual bent (like Arthur Caplan, for instance), of "measuring past and present consent procedures with the yardstick of the autonomous authorization type implies a strong risk of axiological imperialism." In fact a "new concept of consent born inside American law ... has been raised to the condition of a universal model" by means of which an attempt has been made to re-describe the history of informed consent in medical ethics. This kind of "theoretical magnification brings discredit or belittlement on the pre-Nuremberg customs and uses related to consent." Thus, "in front of such a bold self-attribution of historical importance and ideological superiority, it is a duty for all to wonder whether that version of the history of informed consent takes account of all the available data and analyzes them with fairness" (Herranz 2004, 134). I am in agreement with Herranz that the answer to a question thus framed is negative. By way of explanation, it nevertheless must be added that Faden and Beauchamp are not the most radical proponents of the method described by Herranz. Their views, despite clear dependence on the new authorization type of consent, are introduced in a diplomatic manner and tend to be moderate. Nevertheless, for the reasons just provided, their conclusions situating the origins of informed consent only in the twentieth century cannot be considered sufficiently coherent and conclusive.

In my opinion, among the various voices in the debate over the birth of informed consent, that of Pernick seems to present itself as most realistic and believable. In terms of methodology employed, Pernick, a historian, proves much more resistant to any projection of philosophical or clinical assumptions onto the historical data. This renders his approach substantially more consistent than that of Faden and Beauchamp and less emotionally loaded than that of Katz.

The remarks provided by Herranz are also of the utmost relevance for the subject matter. The next section will expand on his perspectives.

1.2 The Roots of the Principle of Free and Informed Consent in the Catholic Tradition Prior to World War II

Faden and Beauchamp believe that "the nineteenth century saw no hint of a rule or practice of informed consent in clinical medicine" (Beauchamp and Faden 2004, 1272). It seems at least in some measure true that informed consent in its current form did not exist before the twentieth century as a concept widely recognized in both medical and legal practice, and that the precise expression "informed consent" was not in use. Still, the idea of informed consent was operative before 1901, as Gonzalo Herranz demonstrates.

In "Some Christian Contributions to the Ethics of Biomedical Research: A Historical Perspective," Herranz claims that a prevailing and commonly accepted account of the ethics of biomedical research (EBMR) in contemporary literature tends to focus selectively on certain events only at the cost of unjustified diminishing or omission of others. Among the overlooked facts of EBMR history are the groundbreaking contributions of Christian ethics.

Herranz makes a credible critique of the following claims, which are easily found in what he calls "the standard history" of EBMR. First, there is the conviction that the Nuremberg Code (1947) was the starting point of the EBMR and that nothing significant took place before its release. The second claim is the belief that it was only after the Declaration of Helsinki (1964) and the Belmont Report (1979) that EBMR became recognizable as a distinct area of medical ethics. Third, the role that secularist ethics played in the process of creating EBMR has been over-emphasized. Moreover, a serious attempt has been made to portray consent which is genuine and fully respectful of patients' autonomy as a product that is purely American (see Herranz 2004, 132–35). Herranz's research demonstrated that these claims are indefensible in the light of Christian achievements made by the nineteenth-century European school dealing with medical ethics.

1.2.1 An Early Claim for Free and Informed Consent

It is of great interest that the obligation of free and informed consent was made clear by Catholic doctors belonging to the nineteenth-century French school of *Morale Médicale* with respect to nontherapeutic research on human subjects. According to Herranz, "in the Catholic tradition, doctors and patients, researchers and subjects, must be guided by a spirit of moral rectitude, brotherly love, sincerity and freedom. All are equally human beings, in the image of God, installed in identical worth, done by divine grace capable of maintaining a direct, personal and filial relationship with God" (Herranz 2004, 135–36). Remarkably, a natural disproportion between doctor and patient or researcher and subject is remedied in the Catholic tradition not by over-emphasis on the patient's or subject's autonomy but by direct recourse to theology that recognizes a fundamental human equality.

A meaningful application of theological premises to medical practice can be found in Georges Surbled, a practitioner of the *Morale Médicale* school. Surbled linked the insensitivity of some researchers and scientists directly to their lack of faith in God. The point here is that the love of neighbor, regardless of his or her state of health and social status, should be rooted in love of God. Believing himself that "charitable love was the first and last word of science" (Surbled 1905, 216–17; quoted in Herranz 2004, 136). Surbled grounded his call for "formal consent" in human rights and the principle of justice. Any research subject "has the right to be respected in his body" and thus he can never "be used as a cheap object for experimentation." Moreover, the subject has a right to reject an experiment or a treatment without giving reasons justifying his decision. These acknowledgments, as Herranz stated, articulate what was subsequently acknowledged in a well-known and repeatedly cited statement made in 1914 by a U.S. judge in regard to informed consent: "every human being of adult years and sound mind has a right to determine what shall be done with his own body; and a surgeon who performs an operation without his patient's consent commits an assault, for which he is liable in damages" (Katz 1972, 526; quoted in Herranz 2004, 137).

It is noteworthy that Surbled easily reconciled his views with the ancient Hippocratic rule "primum non nocere" (first, do no harm), to which he also subscribed. This rule, associated in the current mainstream literature with the principle of beneficence, has been portrayed by some scholars as justifying a paternalistic and authoritarian view of the patient-physician relationship, which diminishes the patient's autonomy (see e.g. Faden, Beauchamp, and King 1986, 10, 37, 62). Surbled's position shows, however, that this is not necessarily the case. Since the patient is "the only master of his body to use it in freedom," any unauthorized use or any use based on misinformation or deception rather than on adequate information by a physician is harmful and as such should result in criminal charges. Surbled explicitly brings out the relevance of information that needs to be given to the subject before securing consent. According to his own words, "the subject, after being *fully informed* of the incurred risks, must lend himself to the research absolutely free and of his own will" (Herranz 2004, 138, emphasis added). As Herranz puts it, "Surbled stated unambiguously that the patient's best interest imposes itself with such strength on the physician's conscience that it determines his behavior" (Herranz 2004, 138). Notably, the subject's best interest is not established by a researcher; it is rather ascertained by the patient himself, although not necessarily without the researcher's advice.

Although "Surbled's concept of informed consent does not yield in ethical depth to the notion of consent enshrined in the more advanced documents of today," there is no doubt it satisfies all the criteria essential for an acceptable form of informed consent as proposed by Faden and Beauchamp and quoted in the last section (Herranz 2004, 137). Consequently, Surbled's work shows that Faden and Beauchamp are incorrect not only with respect to the historical origins of informed consent but also in regard to the role that Christian—or more precisely, given the setting of Surbled's work—Catholic ethics played in medical practice before the twentieth century.

Theologically inspired Catholic moral tradition goes far beyond Faden and Beauchamp's approach and in fact beyond any other ethics redefining human freedom by overemphasizing autonomy, which in turn relegates freedom to an isolated individualistic self-expression. The notion of autonomy critical to the Catholic intellectual heritage implies a set of fundamental truths that has remained unrecognized by the ethics of a solely secular origin. Some of those truths refer explicitly to God while others build chiefly on philosophical premises. Whatever their epistemological character, the principles central to Catholic ethics emphasize that human life and welfare can never be compromised by being used merely as means in the pursuit of new medical knowledge. From a theological perspective, human life, as entrusted by God to humans, remains God's property. It follows that humans should administer their lives as gifts given by God, with responsibility and prudence. In addition, as will become more evident in later chapters of this study, human freedom develops and actualizes itself within a community and by means of a community of persons rather than as the work of an isolated individual. Subsequently, from the Catholic viewpoint, freedom can never be downgraded to mere choice, expressing itself in an arbitrary manner and depending solely on the subject's preferences. As Herranz's research has demonstrated, the achievements of the school of *Morale Médicale* remain consonant with and derivative of this Catholic moral thinking.

1.2.2 The Principle of Superiority of Persons over the Interests of Science and Society

As will be presented below in a more detailed fashion, the 1975 Declaration of Helsinki in its second version (known as Helsinki II) mandates the supremacy of the interests of human individuals over the interests of science and society. This ethical stance was neither new nor novel at the time of Helsinki II's publication. As early as 1845 it was clearly recognized by a French physician, Max Simon, who insisted that "neither the concern for science, nor the decision to solve an important theoretical problem or the craving for adding a new agent to the medical armamentarium, can lead experimenters to lose touch with the immediate interest of the individual who is the subject of their studies. Nothing can release doctors from their humanitarian duty to assure the suffering patient the whole benefits of their art . . . It is not possible to over-emphasize this principle: the most indigent and worthless patient, the more useless to society, cannot be subjected to risky or dangerous experiments. Perish rather science than this principle!" (Simon 1845, 335, 337; quoted in Herranz 2004, 139). These sharp words leave us with no doubt as to Simon's concern about the primacy of human subjects over any scientific end. Like Surbled's position on the importance of concern for the subject's consent, Simon's approach can also be traced back to the underlying principles of Christian moral doctrine. For instance, their ontological and ethical status, implying a unique dignity, never allows the treatment of humans merely as means to other ends.

Simon's apprehension about potential abuses of human dignity in the name of science stands in complete harmony with the address Pope Pius XII delivered to the First International Congress on the Histopathology of the Nervous System, on September 14, 1952. The pope concisely presented the Church's teaching on the matter, asserting, among other things, “that science is not the highest value to which all other values are to be subordinated; that the human person cannot be used as an object by the community; and, finally, that the patient is not the absolute master of himself: he cannot, therefore, freely dispose of himself as he pleases” (Herranz 2004, 140).

Thus, the Magisterium’s teaching on the matter as well as the Declaration of Helsinki were anticipated by a lay Catholic physician, deeply preoccupied with the application of Christian morals to the field of research involving human subjects. In addition, Simon’s concern extended to possible methods of implementing and executing moral principles safeguarding human subjects in concrete research settings. This is the very problem that has found its acute, large-scale expression in the twentieth-century debate over the legal force of human rights and their impact on national legislation and medical practice. We will next consider how the treatment of this issue has developed since the mid-twentieth century.

Chapter 2
The Articulation of the Principle of Free and Informed Consent in Human-Rights Documents

After World War II it became evident that scientific research needed international oversight in order to eliminate or at least to curtail abuses of human subjects. Here I will briefly examine several documents that seek to achieve this goal. The principle of informed consent occupies an essential position in each of them. Yet, given the rapid development of medical science as well as its close connection to law, politics, and demographics in diverse cultural and economical milieus, the position of informed consent may vary from one document to another. What is more, it might be argued that the principle itself, so clearly and sharply articulated in the Nuremberg Code, has been gradually devalued with the passing of time. Regardless of its articulation in human-rights documents and its strong legal status in the United States and in many countries of Western Europe, the principle has also been compromised even to the point of open violation in the post-World War II era.

2.1 The Nuremberg Code

The Nuremberg Code is a set of ten principles included in the final legal verdict delivered at the Nuremberg trial on July 19, 1947. The trial, known legally as *United States of America v. Karl Brandt et al.*, commenced on December 9, 1946, at the Palace of Justice in Nuremberg, Germany. The 23 defendants, Nazi physicians, were charged with war crimes and crimes against humanity, including ruthless, involuntary, predominantly military-related research on humans. In the course of the trial two major claims made by the defendants were rebuffed: first, "the contention that only a state, and not individuals, could be found guilty of war crimes," and second, "the argument that the trial and adjudication were ex post facto" ("Nürnberg Trials" 2008). On account of the relevance of both the Nuremberg Code and its historical predecessors with respect to informed consent it seems desirable to dedicate a separate section to the Code's historical and ethical background. Likewise, an additional section has been provided to describe the role the Code plays in U.S. and international law.

G. Mazur, O.P., *Informed Consent, Proxy Consent, and Catholic Bioethics*, Philosophy and Medicine 112, DOI 10.1007/978-94-007-2196-8_2,

2.1.1 Historical and Ethical Background of the Nuremberg Code

Throughout World War II, Nazi doctors under the pretence of science and medical progress performed extensive experiments on concentration camp prisoners and on prisoners of war.[6] These experiments were often no more than cruel murders with no or little value for medical science. Many of them were carried out only with the goals of killing efficiently and of accelerating death. Regardless of their purpose, the most appalling included high-altitude and low-pressure experiments, freezing experiments, malaria experiments, mustard gas experiments, sea-water experiments, epidemic jaundice, sterilization, and typhus experiments.[7]

As Grodin points out, the Code originated primarily in the minds of two men, one a professor of physiology and pharmacology, Andrew Ivy, and the other a professor of psychiatry and neurology, Leo Alexander (Grodin 1992, 140, note 3). During the trial, they were employed by the prosecution as the chief experts "for history and ethical standards of human experimentation" (Grodin 1992, 122). Since both the prosecution and the defense used in their arguments different ethical guidelines and criteria present in medical science prior to the trial, the references made by Ivy and Alexander appear to be of the highest importance for understanding the origins of the Code.[8] Four sources particularly influenced their thinking and arguments.

The first source is Hippocrates and the Hippocratic tradition of beneficent medicine. According to Ivy, the Hippocratic Oath "represents the Golden Rule of the medical profession. It states how one doctor would like to be treated by another doctor in case he is ill. And in that way how a doctor should treat his patient or experimental subjects. He should treat them as though he were serving as a subject" (Military Tribunal 1 1950; quoted in Grodin 1992, 123). Alexander adds to this that "every professional relationship between the physician and another human being, irrespective of whether the physician treats the patient, examines him or performs an experiment upon him with his permission, is bound by the principles laid down in the Hippocratic Oath" (Alexander 1976, 41; quoted in Grodin 1992, 123).

The second source of Ivy's and Alexander's argumentation is the aforementioned English physician Thomas Percival's *Medical Ethics* and the first American Medical Association Code of Ethics in the United States, published in 1847, which was based on Percival's work (Grodin 1992, 124). Percival's code is considered to be the first to contain guidelines concerning research on human subjects. After stating that "new remedies and new methods of chirurgical treatment should be devised," Percival makes it clear that "the gentlemen of the faculty should be scrupulously and conscientiously governed by sound reason, just analogy, or well-authenticated facts. And no such trials should be instituted without a previous consultation of the physicians or surgeons according to the nature of the case" (Percival 1985, 15; quoted in Grodin 1992, 124). In other words, as Grodin concludes, "the focus ... of this early guide to research ethics is on good methodology and competent investigators. While both are crucial there is no mention ... of the need for the protection of human subjects, nor is there any discussion of consent at all" (Grodin 1992, 125).

Some information on both the safety of the human subject and on the requirement for consent can be found in the work of William Beaumont published in 1833, the first American text treating the topic of experimentation on human subjects. Beaumont was a "U.S. army surgeon, the first person to observe and study human digestion as it occurs in the stomach" ("Beaumont, William" 2008). His guidelines make up the third source used by Ivy and Alexander in the origination of the Nuremberg Code (Grodin 1992, 125).

According to Grodin, the last source used by Ivy and Alexander is the work of the French physician Claude Bernard, who in his 1865 *Introduction to the Study of Experimental Medicine* formulated major principles for the ethics of experimentation on humans. What is of most importance for this discussion is that Bernard distinguished between therapeutic and nontherapeutic research, claiming that only the former has ethical legitimacy. According to him, "the principle of medical and surgical morality ... consists in *never performing* on man an experiment which might be harmful to him *to any extent*, even though the result might be highly advantageous to science" (Bernard 1927, 101–102; quoted in Grodin 1992, 125–26, emphasis added). It would follow, then, that any risk, however minimal, makes potential research unethical, i.e., not to be performed on a human subject. In other words, it is the personal benefit of the subject that validates any research to be performed on him. Bernard not only accepts but also underscores the relevance of and the need for such a research. According to him, "those [experiments] that are innocent are permissible, and those that may do good are obligatory" (Grodin 1992, 126).[9]

As Grodin summarizes the issues, "the codes of Hippocrates, Percival, Beaumont, and Bernard are all concerned with a physician's responsibility to benefit the patient/subject. Hippocrates deals only with the physician-patient relationship, Percival addresses innovative therapies, Beaumont covers nontherapeutic experimentation, and Bernard focuses on the scientific method and therapeutic research. Beaumont and Bernard are also concerned with acceptable experimental risk. Only Beaumont provides any discussion of voluntary consent as a necessity for human experimentation" (Grodin 1992, 126).

The theme of voluntary consent turned out to be more significant in the context of legal regulations characteristic of the pre-World War II German medicine in which the defendants of the Nuremberg trial developed as professional physicians and researchers. While speaking of the historical background of the Nuremberg Code, this context must be remembered along with the above-mentioned codes because of the influence it exerted on Ivy's and Alexander's thinking, as well as on the judges' verdict. One of the major arguments of the defense during the Nuremberg trial can be described as twofold. First, it maintains that in order to understand more fully and to assess correctly the experiments carried out by the defendants during the war, one needs to invoke the codes of medical ethics that were in use in Germany prior to World War II. Second, the defense maintained that the relativism of different ethical guidelines or codes with respect to experimentation on humans makes the codes of "earlier times or other locales" (Grodin 1992, 128) not constitutive of German medical standards and thus not applicable to the experiments carried out by German physicians during the war.

An example of a code of "earlier times" would be a document concerning the ethical bounds of nontherapeutic experimentation on humans, issued by the Prussian Minister of Religious, Educational, and Medical Affairs on December 29, 1900. The text states that "medical interventions for purposes other than diagnosis, therapy and immunization are absolutely prohibited" with respect to minor and "not fully competent" subjects. They are also outlawed when "the unequivocally declared" consent is lacking or when misinformation rather than "a proper explanation of the adverse consequences that may result from the intervention" takes place prior to the experiment. Moreover, the document provides significant procedural limitations in regard to the performance of nontherapeutic experiments. They can be carried out "only by the director of the institution himself or with his special authorization." In addition, a "medical case-record book" needs to be kept in order to verify if all the requirements have been met in regard to a concrete case (after Grodin 1992, 127).

Taking into account the time of its origin, the Prussian document seems to be a remarkably mature text seriously aiming at the full protection of human subjects of nontherapeutic experimentation. Significantly, it refers favorably to the notion of consent that should be informed and "unequivocally declared." Although there is no ultimate certainty as to the meaning of the combination of these last two words, it can be reasonably assumed that consent thus described must be given in a free manner. Furthermore, as Grodin observes, the document being discussed "may be the first dealing with the ethics of human experimentation that specifically recognizes the need for the protection of uniquely vulnerable populations such as minors or incompetents" (Grodin 1992, 127).

In the light of the Prussian text one may more easily understand why the Nuremberg defense tried to exclude it as anachronistic in establishing what had been the relevant "standard for Germany." There is no doubt that if the judgment on the Nazi experiments was made on the basis of the Prussian directive only, the defendants could easily have been found guilty. Nonetheless, even if one respects the defense's demand and ignores "codes of earlier times," there is still at least one other document that originated in Germany after 1930 and that sheds valuable light on the matter. It is unquestionable that the highly relevant "Regulations on New Therapy and Human Experimentation," promulgated in 1931 and cited by Ivy during the trial, contain almost all of the provisions later included in the Nuremberg Code. Since the document presents a great variety of issues concerning research on human subjects, including informed consent, and does it in a very mature manner, some regard it to be in numerous aspects even more far-reaching than both the Nuremberg Code and the Declaration of Helsinki (Grodin 1992, 132). At the same time, however, as Hans-Martin Sass notes, "the fact that such a governmental regulation actually existed, at a time when the Nazis were carrying out human experiments in an irresponsible manner in concentration camps, underlines the irrelevance of legal regulations if they are not enforced by the authorities" (Sass 1988, 52).

It is no wonder then that there was a great deal of controversy during the trial over the Regulations' validity and legal force. Again, the defense made an effort to get around the guidelines, this time by pointing to the fact that they were not officially recognized by the Office of Public Hygiene, the international institution designed to oversee both the national and the international legal regulations regarding health. Nevertheless, the Regulations were found by the Nuremberg tribunal to be a paradigm of ethics dealing with experimentation on humans during the Nazi era in Germany (see Grodin 1992, 129).

At the end of the trial, Ivy and Alexander decided to summarize their position based on the sources already mentioned for the purposes of creating a scheme of the principal ethical points concerning experimentation on humans and the requirement of informed consent, which could be employed by the judges in the process of seeking the final verdict. After several modifications, the proposal was accepted by the tribunal and included in the final judgment in the form of ten points to be subsequently called the Nuremberg Code. In the next section I will briefly present the Code's content.

2.1.2 Content of the Nuremberg Code

In its very first article the Nuremberg Code affirms that the "voluntary consent of the human subject" is absolutely essential. "This means that the person involved should have legal capacity to give consent; should be so situated so as to be able to exercise free power of choice, without the intervention of any element of force, fraud, deceit, duress, over-reaching, or other ulterior form of constraint or coercion; and should have sufficient knowledge and comprehension of the elements of the subject matter involved, as to enable him to make an understanding and enlightened decision. This latter element requires that, before the acceptance of an affirmative decision by the experimental subject, there should be made known to him the nature, duration, and purpose of the experiment; the method and means by which it is to be conducted; all inconveniences and hazards reasonably to be expected; and the effects upon his health or person, which may possibly come from his participation in the experiment."[10]

The Nuremberg Code also states that experiments carried out on humans should be "unprocurable by other methods or means or study . . . not random and unnecessary in nature" (art. 2) and "so designed and based on the results of animal experimentation" or knowledge derived from other studies "that the anticipated results will justify the performance of the experiment" (art. 3). Moreover, not only should "no experiment . . . be conducted where there is an a priori reason to believe that death or disabling injury will occur" (art. 5), but also "the experiment should be so conducted as to avoid all unnecessary physical and mental suffering and injury" (art. 4). The related article 7 goes further, stating that "proper preparations should be made and adequate facilities provided to protect the experimental subject against even remote possibilities of injury, disability, or death."

The Code puts certain limitations on the risk that is acceptable during experimentation. Article 6 reads: "the degree of risk to be taken should never exceed that determined by the humanitarian importance of the problem to be solved by the experiment." In other words, it is not the scientific issue but the human subject that remains of the highest importance during the course of the experimentation. This conclusion anticipates the further legal regulations to be examined in this chapter. At the moment I want to mention only the problem of what is called "significant risk" with respect to any experimental conduct. The Code, for understandable reasons, does not specify what is meant by the appropriate degree of risk to be taken. This degree varies greatly depending on the concrete research situation. It needs to be explored and assessed by both researcher and subject by means of dialogue preceding the securing of consent.

The emphasis on the central place that every person undergoing experimentation takes in the research practice has been confirmed by the Code in regard to human freedom and to what is now often called, especially in the legal context, "the right to self-determination." "During the course of the experiment, the human subject should be at liberty to bring the experiment to an end, if he has reached the physical or mental state where continuation of the experiment seemes to him to be impossible" (art. 9). The same remains true on the part of the researcher who "must be prepared to terminate the experiment at any stage, if he has probable cause to believe, in the exercise of the good faith, superior skill and careful judgment required of him, that a continuation of the experiment is likely to result in injury, disability, or death to the experimental subject" (art. 10). In these words the Code expresses a very fundamental concern about the hierarchy of values to be adopted by any researcher with respect to the research subject: it is not the good of experiment or any other widely understood good of science, but of the person who is the subject, that takes priority in experimentation on human subjects.

According to the analysis of the Nuremberg Code offered by Leonard Glantz, the entire Code can be divided into two sections. The first consists of articles 9 and 1 whereas the second contains the remaining articles. Taken together, clauses 9 and 1 create a basic security framework for every single person potentially involved with research as a subject. As Glantz observes, "these provisions protect the rights of the individual as an autonomous human being. Individuals are given the absolute right to refuse to be subjects and the absolute right to terminate their participation at any time. In a sense, human subjects as a group are given 'veto power' over research," that power of which the victims of the heinous World War II experiments were brutally deprived. The remaining eight articles do not handle the subject's decision concerning participation in research but rather the conduct of researchers who are either prohibited from carrying out selected experiments or are required to safeguard those who consented to be engaged in them. "Thus, 8 of the Code's 10 provisions are directed not at protecting the rights of subjects but rather at protecting their welfare. These provisions focus on researchers and instruct them what things may not be done and what things must be done regardless of the subjects' consent" (Glantz 1992, 184). To put it another way, the Code aims at providing an objective standard for experimentation on humans: the standard is independent of the subjective perspectives of both researcher and research subject. On one hand the Code defends the subject's autonomy; on the other, however, it highlights the limited nature of that autonomy. Thus, *Nuremberg* opts for what can be called the limited notion of autonomy, as opposed to its absolute or close-to-absolute versions.[11] Indeed, the consent, even if free and legitimately secured, does not absolve the researcher of responsibility for carrying out experimentation. Consequently, when great risk to the health or life of the subject comes into play, the researcher, in the name of responsibility, should not even initiate the research, even when informed consent has been validly obtained.

Among other characteristics of the Nuremberg Code offered by Glantz, the following two, I believe, are most important. First, the Code addresses its requirements in a direct line to the researcher and not to "the institution that sponsors the research or in which the research is conducted. The rules embodied in the code are self-executing and do not require the involvement of any review board or similar bureaucratic entity." Second, given the backdrop of the Code as well as the process of its coming into being, it seems reasonable to assume that its obligations relate to nontherapeutic rather than therapeutic research. This inference finds its confirmation in the fact that "*none* of the Code's provisions refer to balancing the risks of the research against the benefits the subject might be expected to obtain from participation." Yet, if it is true that the Nuremberg document regulates pure experimentation, that is, experiments unrelated to any potential profit of the subject, then it "is directed at a relatively simple problem and has only limited relevance to the complexities of research with human subjects as it is conducted today" (Glantz 1992, 185).

2.1.3 *Influence of the Nuremberg Code on International and U.S. Regulations*

There is no doubt that the Nuremberg Code is a significant attempt at providing a standard based on natural law for the ethics of experimentation on human subjects. As the preface to the Code states, "*all agree* ... that certain basic principles must be observed in order to satisfy moral, ethical and legal concepts" (emphasis added). This formulation clearly points to the existence of fundamental rationales for the protection of human beings which are accepted by all. In other words, the Code is meant to be a natural-law document because it originated out of "the principles of the law of nations as they result from the usages established among civilized peoples, from the laws of humanity, and from the dictates of public conscience" (Military Tribunal 1 1950, 183; quoted in Perley et al. 1992, 152).

As has been stated, the Code represents "the most complete and authoritative statement of the law of informed consent to human experimentation" (Annas, Glantz, and Katz 1977, 1). The universal perspective taken by the judges in the process of formulating the final verdict was widely appreciated after the trial ended.[12] The Nuremberg tribunal "gave the first impetus to the worldwide movement for international human rights ... The Universal Declaration of Human Rights, agreed to on December 10, 1948, is quoted and used in the constitutions of the more than 100 new nations that rose from the ashes of colonialism after World War II." In addition, "22 covenants on human rights have emerged from the United Nations, at least partially due to Nuremberg. There are covenants against racism, sexism, torture, and genocide" (Drinan 1992, 178). The fact that the Code "has been embodied in many, if not all, subsequent ethical codes governing biomedical research involving the use of human subjects" (Perley et al. 1992, 152), firmly confirms its influential character. As an example one can invoke the regulations found in such early documents as the International Covenant on Civil and Political Rights, the Declaration of Geneva (considered to be a contemporary reaffirmation of the principles included in the Hippocratic Oath), and other World Medical Association documents like the 1954 Resolution on Human Experimentation (Perley et al. 1992, 153–55). Similarly, the 1964 Declaration of Helsinki reveals a solid connection to the Nuremberg Code that is to be presented in a more detailed manner in the next section. In summary, at the level of international law the role that the Code has played remains substantial.

Nonetheless, some authors claim that despite the fact that the Nuremberg Code presents the international community with the regulations critical to the ethics of experimentation on humans, it remains primarily a historical document with no significant impact on contemporary areas of research (Howard-Jones 1982, 1443).[13] Surprisingly, since the Code's coming into existence its provisions have been compromised and even directly violated, as was the case, for instance, with the informed consent waiver granted in 1990 by the United States Food and Drug Administration at the request of the U.S. Department of Defense concerning the use of unapproved drugs and vaccines on the soldiers engaged in Operation Desert Shield (see "Department of Defense Request . . ." 1990).[14]

Certainly, while the Code has been acknowledged in the international forum, greater complexity has arisen with regard to U.S. law. The Code itself "is part of international common law and may be applied to both civil and criminal cases, by state, federal and municipal courts in the United States" (Annas, Glantz, and Katz 1977, 21). Despite that fact, the Code has paradoxically never been employed in any criminal case and it was cited only a few times in the civil cases by a U.S. court. Given the origins of the Code, as a document that resulted from a trial "conducted by U.S. judges under the authority of the U.S. military," where "U.S. procedures were followed and U.S. lawyers acted as prosecutors" (Annas 1992, 201), this conclusion is remarkable. Indeed, it is even more striking when one realizes that the first U.S. court cited the Code only in 1973. Annas offers two reasons as feasible explanations for this fact. First, it is noteworthy that the number of legal cases concerning experimentation was quite small at that time. Second, both jurisprudence and medicine claimed no connection whatsoever to the barbarity of the victimization of research subjects in the course of World War II. The Code, the origin of which was so intensely marked by the atrocities of war, was thought of as overly stringent and therefore inappropriate for guiding the performance of experiments in peacetime. Interestingly, this mindset contributed in a prominent way to the development of the Declaration of Helsinki.

2.2 Declaration of Helsinki

As suggested in the previous section, the Nuremberg Code, although relevant, required improvement and development. Many experts found the need to create "professional guidelines designed by physicians for physicians (as opposed to the

Nuremberg Code, which was formed by jurists for use in a legal trial)" (Perley et al. 1992, 157). In the U.S., Henry Beecher proposed in 1960 that a new international document "might be developed which would serve the best interest of science and mankind in the realm of human experimentation" (Beecher 1960, 80; quoted in Perley et al. 1992, 157). One of the most urgent demands in this realm was to classify and work out different types of research, particularly those concerning diagnostic and therapeutic practices as opposed to those aimed at other goals.

2.2.1 Helsinki I

In response to these challenges the World Medical Association (WMA) accepted in 1954 a five-principle "Resolution on Human Experimentation: Principles for Those in Research and Experimentation." Then, as a result of further studies, debates, and revisions, a new code was adopted by the 18th World Medical Association Assembly, in Helsinki, Finland, in June 1964. The code was christened the Declaration of Helsinki. Although nowhere explicitly mentioned in the document, the Nuremberg Code served as a basic point of reference for the drafters and was discussed at WMA meetings. Without doubt the two documents remain in harmony as to the main points concerning human experimentation. Like the Nuremberg Code, the 1964 Declaration of Helsinki (Helsinki I) makes it clear that clinical research on humans "must conform to the moral and scientific principles that justify medical research, and should be based on laboratory or animal experiments or other scientifically established facts"[15] (principle I.1). Moreover, as Sharon Perley et al. say in summary, "both documents require that research be conducted only by scientifically qualified persons ... Both provide that at any time during the experiment, the research subject should be free to withdraw. Also, both provide that at any time during the experiment, the investigator should discontinue the research, if, in his judgment, its continuation would be harmful to the subject ... Finally, both the Nuremberg Code and Helsinki I require that the researcher obtain informed consent" (Perley et al. 1992, 158).[16]

Along with similarities to the Nuremberg Code found in the 1964 Declaration of Helsinki, there are also certain differences, although these are complementary to the Nuremberg Code.[17] Two of them deal directly with informed consent.

The first, regarded by many as the greatest supplement to the Code, is Helsinki I's distinction between therapeutic and nontherapeutic research. Even in its introductory section, the Declaration asserts that "a fundamental distinction must be recognized between clinical research in which the aim is essentially diagnostic or therapeutic for a patient and clinical research the essential object of which is purely scientific and without therapeutic value to the person subjected to the research." Perley et al. conclude that "the Declaration thus recognizes that ethical principles that must be followed when experimenting on healthy volunteers must also be followed when experimentation is combined with therapeutic care" (Perley et al. 1992, 158).

The second difference is that Helsinki I, unlike the Nuremberg Code, allows for the possibility of securing informed consent from a legal proxy in both therapeutic and nontherapeutic experimentation provided that a subject remains unable to give consent. Article II.1 reads that "in case of legal incapacity consent should also be procured from the legal guardian; in case of physical incapacity the permission of the legal guardian replaces that of the patient." This assertion remains in clear contrast with the Nuremberg Code, which does not remark on physical capacity and explicitly states in the first article that "the person involved *should have* legal capacity." Despite this contrast, however, the two documents may be, at least to a degree, complementary to each other with regard to the legal capacity requirement. In truth, the Helsinki I provision can be viewed as qualifying the Nuremberg Code's requirement by making reference to legal guardians, which points to the fact that a patient who cannot make a decision due to physical incapacity still has the legal capacity to do so through an authorized proxy.

Third, Helsinki I adds an important consent-related note to what was stated in the Nuremberg Code. "Consent—as principle III.3c puts it—should as a rule be obtained in writing."[18] In addition, securing consent does not absolve a researcher from responsibility for experimentation. This responsibility, in the words of the principle, "always remains with the research worker" and "never falls on the subject, even after consent is obtained." Moreover, unlike the Nuremberg Code, Helsinki I demands not only that "clinical research should" be done "by scientifically qualified persons" but also ought to remain "under the supervision of a qualified medical man" (principle I.2).

2.2.2 Helsinki II

In the face of swift advances in medical science, the Declaration of Helsinki has been revised and amended several times.[19] The first, and arguably most important, update to Helsinki I took place in 1975 in Tokyo and has been labeled Helsinki II. As Perley et al. state: "Helsinki II is recognized by most as providing the fundamental guiding principles for the conduct of biomedical research involving human subjects. It has been adopted, in modified but similar forms, in international texts and national legislation and by many professional medical organizations throughout the world" (Perley et al. 1992, 159). In fact, certain scholars still regard Helsinki II as the most representative version of the Declaration and consider the changes introduced by its later revisions to be minor (Jonsen, Veatch, and Walters 1998, 13). In what follows I will take a closer look at major modifications adopted by Helsinki II with respect to Helsinki I and the Nuremberg Code. All of them directly or indirectly address and affect the issue of informed consent.

First, Helsinki II, despite keeping the distinction between therapeutic and nontherapeutic research, alters the terminology from what was previously called "clinical research" to "medical research combined with professional care" and from "non-clinical biomedical research" to "nontherapeutic biomedical research involving human subjects." This change qualifies and better corresponds to what actually takes place in the area of medical research and thereby sets a new standard of terminology widely accepted in the professional literature (Perley et al. 1992, 160).

Second, Helsinki II asserts in principle I.5 that "concern for the interest of the subject must always prevail over the interest of science and society." The same is repeated in the third section, dedicated to nontherapeutic biomedical research involving human subjects: "in research on man, the interest of science and society should never take precedence over considerations related to the wellbeing of the subject" (principle III.4). This reiteration is remarkable and telling. Researchers must never succumb to the temptation to compromise their human subjects' dignity. If they should do so, this almost always strikes at both the care of the subject and the way his or her informed consent is secured. Well-documented instances exist in which researchers' interest in science, extensive desire for profit or fame, or inexcusable negligence have overcome respect for the research subject.[20] As history teaches, different kinds of manipulation and coercion can easily replace a genuine consent with a fraudulent substitute which compromises the subject's dignity and wellbeing.

Third, Helsinki II highlights the need for independent supervision over research, stating that "the design and performance of each experimental procedure involving human subjects should be clearly formulated in an experimental protocol which should be transmitted to a specially appointed independent committee for consideration, comment and guidance" (principle I.2). In addition, as principle I.12 has it, "the research protocol should always contain a statement of the ethical considerations involved and should indicate that the principles enunciated in the present Declaration are complied with." The requirement of independent ethical review boards is regarded by some authors as the most significant update to Helsinki I (see Perley et al. 1992, 159). It was also developed and improved by subsequent versions of the Declaration.[21] Clearly, the requirement dealing with research protocols and review boards has profound consequences for informed consent, as it aspires to ensure high standards for obtaining the consent. As is widely admitted, the signature of a patient or subject on a consent form does not suffice to prove a free and adequately informed consent.

Fourth, unlike Helsinki I, Helsinki II places the requirement of informed consent among the basic principles, thus giving it an even greater importance. Principle I.9, for instance, reads not only that "each potential subject must be adequately informed of the aims, methods, anticipated benefits and potential hazards of the study and the discomfort it may entail" but also—a statement that stands in a full agreement with the Nuremberg Code—that a subject has the right to "abstain from participation in the study and . . . is free to withdraw consent to participation at any time." Moreover, "when obtaining informed consent for the research project, the doctor should be particularly cautious if the subject is in a dependent relationship to him or her or may consent under duress. In that case the informed consent should be obtained by a doctor who is not engaged in the investigation and who is completely independent of this official relationship" (principle I.10).

Finally, despite its emphasis on the requirement of informed consent, Helsinki II nonetheless takes a more relaxed stand in some respects as compared to both the Nuremberg Code and, although to a lesser degree, Helsinki I. An illustration of this can be found in the third section of the Declaration, dedicated to medical research combined with professional care. Principle II.5 declares that "if the doctor considers it essential not to obtain informed consent, the specific reasons for this proposal should be stated in the experimental protocol for transmission to the independent committee." Even if the protocol needs to be approved by the committee before the doctor has a right to waive the consent requirement, the principle does not qualify what kind of reason may justify departure from this requirement, nor does it indicate the criteria that are to be met by the potential members of "the independent committee." Regrettably, the general scope and ambiguous character of the 1975 Declaration's statement leave much room for interpretation in this regard. As will be shown in the next chapter, general agreement exists regarding the fact that a practitioner should not be required to secure informed consent in extreme situations when for practical reasons this is impossible. Yet, since Helsinki II does not specifically address this fact, one may have the impression that in certain situations a consent waiver would be justified only if the committee found it reasonable.[22]

2.2.3 *Helsinki III, IV and V*

In reviewing the Declaration of Helsinki's content with respect to the issue of informed consent, several further modifications adopted in its more recent versions must be mentioned. Although there were no substantive changes made in the texts of the 1983 and 1989 versions of the document (that is, Helsinki III and Helsinki IV), two of their provisions deserve attention. Both of these belong to the section dedicated to medical research combined with professional care and both had already been included in that section by Helsinki II. The first is paragraph II.2, which reads: "the potential benefits, hazards and discomfort of a new method should be weighted against the advantages of the best current diagnostic and therapeutic methods." The second is clause II.3, which appends to the previous one the statement that "in any medical study, every patient—including those of a control group, if any—should be assured of the best proven diagnostic and therapeutic method."

By way of explanation, control groups have been used in clinical research since the mid-twentieth century. The method requires two groups of patients, one of which is provided with some type of treatment while the other is not. With a test of a new drug as an example, "one group receives the drug, the other a product identical in appearance, but which is known to be inert—a so-called placebo.[23] At the end of the trial, the results of which can be assessed in various ways, it can be determined whether or not the drug is effective and safe" ("Medicine" 2008).

Notably, the assumption made by paragraphs II.2 and II.3 is that every research enrollee should be treated by best available methods even when a part of a control group. Consequently, the placebo-based trial, in its classic sense, cannot take place: it always implies that the members of the control group are not treated at all, which entails deprivation of access to the best therapeutic methods currently available. Granting this, however, there is still one question that must be answered. What if no established methods exist? Is a placebo-based trial legitimate in that case? In order to respond to this question as well as to allay doubts about the understanding of both paragraphs, the 1996 revision of the Declaration, implemented by the 48th WMA General Assembly that met in Somerset West, South Africa (Helsinki V), supplemented provision II.3 with the following: "this does not exclude the use of inert placebo in studies where no proven diagnostic or therapeutic method exists."[24] In other words, if any proven prophylactic, diagnostic, or therapeutic methods exist, a new method should be tested against them, thus excluding the prospect of using placebos. The statement also implies that "local circumstances—sociopolitical, financial, infrastructural, cultural—can never justify failure to use the best known drugs or technologies in the control arm" (see Tollman 2001, 1417). This is a powerful claim, especially with respect to developing countries in which access to new methods might be dramatically restricted. Put briefly, "interventions that could be expected to be made available in the United States might be well beyond the financial resources of a developing country or exceed the capacity of its health care infrastructure" (Varmus and Satcher 1997, 1004). Because of this, in many cases those researchers who complied with the Declaration's ban on placebo-controlled study found themselves forced to cease conducting any research. On the other hand, however, the reasoning employed by clause II.3 seems logical and justified. It aims at protecting the research subjects in developing countries in two specific ways. The first is to curb barely-controllable activities of some profit-driven pharmaceutical companies performing drug tests on vulnerable populations, and the second is to guarantee the application of one consistent ethical benchmark throughout the world regardless of the indigenous research milieu.[25]

Such reasoning is highly relevant, particularly in regard to the trials carried out in the Third World (for the most part in Africa and Asia) with the aim of preventing the transmission of HIV from mothers to their infants.[26] These trials, although valuable and promising from the scientific standpoint, were nonetheless controversial from the ethical point of view (see e.g. Annas and Grodin 1998). Much of the controversy involves informed consent. Some commentators hold, however, that the main point under discussion "does not hinge on informed consent, which all the trials have obtained" but rather "on whether it is ethical to test interventions against a placebo control when an effective intervention is in use elsewhere in the world" (Varmus and Satcher 1997, 1004). In other words, is it ethically permissible to expose infants to contracting HIV infection from their mothers who are in control groups in research trials even when there are effective drugs known in developed countries and proven to considerably diminish the incidence of HIV infection?[27] It is worth noting that costs do not play a decisive role in this case. Even costs that far exceed the resources currently available in a given local setting can be covered by wealthy countries, as is often the case with controversial placebo-based research sponsored from abroad.[28]

In my opinion both informed consent and placebo-controlled trials raise serious ethical considerations with regard to HIV trials. It is well known that the reality of obtaining informed consent in Third World countries leaves, to put it mildly, a great deal to be desired.[29] Therefore it is not sufficient to say that since consents were obtained this is no longer an issue. In point of fact, subjects involved with research often are not thoroughly informed as to the true nature of the experiments to be performed, nor do they understand the real degree of risk attached to them (Fitzgerald et al. 2002; see also Pence 2004, 297–99). Many investigators believe that research standards inferior to those applied by developed countries can be validly employed in developing countries.[30] Well-documented studies have been carried out in the course of which human subjects were not treated but merely observed by researchers. In addition, withdrawal of information was commonly practiced (see Angell 2000, 15). Thus, aside from the consent problem, there is another issue, that of truth-telling and of exposing people to more than minimal risk because of the lack of treatment.

2.2.4 Helsinki VI and Notes of Clarification

Of course, Helsinki V almost immediately became the subject of heated debate. Only one year after its adoption, during the 49th WMA Assembly in Hamburg (November 1997), the delegates of the American Medical Association (AMA) presented a proposal for revisions to the Declaration which would appreciably relax some of the current regulations. This, in turn, sparked an animated discussion among medical professionals, ethicists, lawyers, and theologians. Not surprisingly, one of the three key points of the AMA proposal was the employment of placebos in biomedical research, the other two being (1) the distinction between therapeutic and nontherapeutic research (is it really necessary to keep it in the text of the Declaration?) and (2) the notion of standard of care (what is ethically required for participants in biomedical research?) (Human and Fluss 2001). Two particularly influential articles were those of Brennan and Levine (Brennan 1999; Levine 1999). The former argued against the proposed revisions, disclosing their utilitarian character and a profit-driven double standard threatening the dignity and integrity of human subjects in vulnerable populations. The latter presented the need for changing the Declaration's text on account of its distinction between therapeutic and nontherapeutic research, which Levine referred to as spurious. In addition, Levine charged *Helsinki* with including "several provisions that are seriously out of touch with contemporary ethical thinking," as a result of which "many researchers routinely violate the requirements of the Declaration" and thus "rob the declaration of its credibility" (Levine 1999, 531).

After three years of discussions, the 52nd WMA General Assembly, meeting in Edinburgh, Scotland, in October 2000, endorsed a new reformed text of the Declaration (Helsinki VI). Perhaps the most significant change is paragraph 30, which reads: "At the conclusion of the study, every patient entered into the study should be assured of access to the best proven prophylactic, diagnostic and therapeutic methods identified by the study." In addition, clause 29, combining the above-mentioned paragraphs marked as II.2 and II.3 in the previous versions of the Declaration, reaffirms that the new methods "should be tested against those of the best current . . . methods" and that placebos can be utilized in research only where no reliable alternative exists.

As might have been expected, by reinforcing its stand on the use of placebos in clinical trials and by calling for availability of the benefits identified by research to participant subjects, the WMA not only did not appease conflicts around the Declaration, but indeed, contributed to their escalation. In many respects the ethical debate over intricacies of research with humans gained momentum after Helsinki VI. The main focus of this debate remained largely on the validity of the ban on the employment of placebos when proven alternatives to the agent being tested exist. Such a ban, it was argued, would severely fetter scientific progress and thus also prevent many from obtaining new, more efficacious medicines (see e.g. European Agency for the Evaluation of Medical Products 2001; Pharmaceutical Research and Manufacturers of America 2001). In response to these objections, utilitarian in character, and desiring that "no good ethical research is restricted,"[31] the WMA relaxed its stand on placebos by issuing, during its 2002 General Assembly in Washington, a Note of Clarification on Paragraph 29 that was added to the official text of the Declaration. The Note acknowledges that "a placebo-controlled trial may be ethically acceptable, even if proven therapy is available," provided that the following two conditions are met. First, "where for compelling and scientifically sound methodological reasons its use is necessary to determine the efficacy or safety of a prophylactic, diagnostic or therapeutic method." Second, "where a prophylactic, diagnostic or therapeutic method is being investigated for a minor condition and the patients who receive placebo will not be subject to any additional risk of serious or irreversible harm."[32]

In the wake of the WMA's acquiescence to heightening persuasion and pressure on the use of placebos, an indirect threat to informed consent arises. As already mentioned, the procurement of valid, i.e., sufficiently free and adequately informed, consent in the Third World is often severely compromised. Now, with a lesser degree of protection for the subjects participating in placebo-control trials, the consent requirement will become even easier to diminish or even to bypass entirely.

Yet, despite its concession on placebos, the WMA retained the requirement of making "prophylactic, diagnostic and therapeutic procedures indentified as beneficial in the study" available to the subjects by issuing a Note of Clarification on Paragraph 30 during the General Assembly in Tokyo in the year 2004. The Note underscores that "post-trial access arrangements or other care must be described in the study protocol so the ethical review committee may consider such arrangements during its review."[33]

The last subject to be covered in this section is the Declaration's legal authority. As is the case with the Nuremberg Code, the Declaration of Helsinki exerts no binding authority on national criminal and civil laws. Perley et al. conclude that these "two sets of principles are just that: sets of ethical principles. Although they are highly influential, neither the Nuremberg Code nor Declaration of Helsinki has any legally binding authority. The rules they set out are tenuous, unaccompanied by any real controls, traditional sanctions, or other means of enforcement. Moreover, as a series of general statements they are ambiguous with respect to both the principles themselves and their practical application" (Perley et al. 1992, 160). From this it follows that the requirements of these codes of ethics are merely the raw material with which particular legislation, usually on a national level and with accompanying enforcement capability, can effectively work. This work always presumes a specific method of interpretation applied to concrete principles. As Jay Katz astutely observed, sole reliance on the codes of ethics would be dangerous, because in fact "they are deceptive documents to which all of us probably could subscribe in principle," but which studied thoroughly, "are painfully vague" (Council for International Organizations of Medical Sciences 1974, 247; quoted in Annas 1992, 205–206).

The truth of Katz's observation notwithstanding, it must also be acknowledged that the implementation of the principles of the both the Nuremberg Code and the Declaration of Helsinki into local legislative systems has been strongly encouraged, frequently with success.[34] While the Nuremberg Code "seemed to have little or no influence on the actual conduct of research, even in the United States, the nation that instigated its promulgation," publication of the Declaration of Helsinki "effected change almost immediately." Consequently, the researchers "were obliged to sign statements that their work had been constructed in accord with the Declaration of Helsinki." What is more, "similar statements were required of authors when they submitted reports of their work for publication" (Levine 1996, 242–43).

These facts, however, do not negate the necessity for further clarification and refinement of the Declaration. Its sound and solid interpretation, as is the case with other codes of ethics, remains of the highest relevance. An attempt to achieve this goal has been made in the CIOMS guidelines, which are to be presented in the following section.

2.3 CIOMS/WHO International Ethical Guidelines for Biomedical Research Involving Human Subjects

A further development in the history of standards for informed consent was the publication by the Council of the International Organizations of the Medical Sciences (CIOMS) of a set of guidelines regarding research on human subjects. The Council "is an international, nongovernmental organization in official relations with the World Health Organization (WHO). It was founded under the auspices of WHO and the United Nations Educational, Scientific and Cultural Organization (UNESCO) in 1949. Among its mandates is that of maintaining collaborative relations with the United Nations and with its specialized agencies, particularly UNESCO and WHO."[35] The other objectives of CIOMS center on (1) facilitating and endorsing international undertakings in the realm of biomedical sciences, and (2) serving the scientific needs of the biomedical community as a whole throughout the world.

2.3.1 Brief Historical and Cultural Introduction to the Guidelines

It was not until late 1970s that CIOMS began to deal with ethical issues arising in the field of biomedical research with human subjects. As new member states of

WHO found themselves in the process of setting up health care systems and ethical review policies after attaining independence, a major concern emerged as to how all the provisions of the Declaration of Helsinki could be successfully applied to the economic and cultural realities of those countries. In 1978 CIOMS, in conjunction with WHO, decided to support developing countries in that process and initiated studies intended for the evolution of ethical guidelines dealing with research on human subjects.

The fruit of these efforts, the 1982 *Proposed International Ethical Guidelines for Biomedical Research Involving Human Subjects*, although useful and valuable, shortly became outdated as a result of the eruption of the AIDS/HIV epidemic and subsequent plans to perform extensive research on vaccines and medicines to prevent and treat it. Coupled with swift advances in biomedical sciences, the need for this kind of research gave birth to new ethical concerns that were subsequently addressed in the 1991 *International Guidelines for Ethical Review of Epidemiological Studies* and the 1993 *International Ethical Guidelines for Biomedical Research Involving Human Subjects.* The latest version of the Guidelines, published in 2002, strives in turn to settle the dilemmas regarding the use of placebos in controlled trials. Although more detailed, the provisions contained in the current version of the Guidelines resemble those found in Helsinki VI and the Note of Clarification on Paragraph 29 discussed in the previous section.

2.3.2 Content of the Guidelines

The document consists of an introduction, an overview of the international instruments and guidelines, the general ethical principles, and a preamble followed by twenty-one guidelines. It addresses a wide variety of issues including the participation of vulnerable populations in research (with a key focus on children, pregnant and nursing women, and the mentally challenged) along with the problems related to social and ethnic diversity worldwide with an emphasis on communal cultures. As Perley has noted, the Guidelines constitute "the first international set of principles to consider the problems associated with externally funded research" (Perley et al. 1992, 164) as well as "the first international code to consider the subject of compensation of research subjects for injury sustained as a result of participation" (Perley et al. 1992, 165).

In the process of creating the Guidelines as well as in their very content, the issue of informed consent occupies a place of distinction. Perley et al. suggested that possibly "the most significant contribution of the WHO/CIOMS Guidelines is the thorough analysis of the problems associated with informed consent" (Perley et al. 1992, 162). The analysis begins by providing in the commentary for Guideline 4 a concise definition according to which "informed consent is a decision to participate in research, taken by a competent individual who has received the necessary information; who has adequately understood the information; and who, after considering the information, has arrived at a decision without having been subjected to coercion, undue influence or inducement, or intimidation." On the basis of this characterization, the four principal elements of consent can be identified. First, the competence of the prospective subject; second, the disclosure of "necessary information;" third, the understanding on the part of the potential subject of the information received; fourth, a free decision made by a potential subject after the consideration of the information.

2.3.2.1 Competence of the Subject

Regarding the first element, the Guidelines do not strictly specify the criteria for competence, leaving this task for national legislatures. In general, every individual of sound mind is capable of giving valid consent. In the case of those lacking competence, the authorization of a legal guardian is required. Guideline 9 provides that (1) ethical and scientific reasons for enrolling incompetent persons in research must be spelled out (it must be proven that such a class of subjects is indispensable for a given type of research, which is to say, less vulnerable subjects cannot be enrolled instead) and (2) the risk inherent to interventions that do not benefit the subject "should be no more likely and not greater than the risk attached to routine medical or psychological examination of such persons." The degree of risk when satisfying this criterion has been depicted in the Guideline 4 commentary as minimal.

Nonetheless, in certain circumstances Guideline 9 allows for more than a minimal risk. In what is called the "low-risk standard," the presence of "slight or minor increases" above minimal risk "might be permitted when there is an overriding scientific or medical rationale for such increases and when an ethical review committee has approved them." On account of the great vulnerability of subjects who are not capable of consent on their own as well as the sometimes less-than-pure motives of their legal representatives, the low-risk standard seems too lenient. Indeed, consent offered by proxies can, on occasion, turn out to be solely profit-driven or to be abusive in a number of other ways. Guideline 7 strives to curb this by providing that since "incompetent persons may be vulnerable to exploitation for financial gain by guardians …, a guardian … should be offered no recompense other than a refund of travel and related expenses." Still, this provision does not change the fact that since the low-risk standard rather than the minimal-risk standard has been adopted by the Guidelines, abuses other than financial ones may take place with respect to incompetent subjects and their unavoidable dependence on legal proxies.

Even so, it needs to be admitted that the minimal-risk standard has not been entirely abandoned by the Guidelines. In fact, it was employed as a major criterion in regard to the issues relating to consent waiver. After stating that "for all biomedical research involving humans the investigator must obtain the voluntary informed consent of the prospective subject or, in the case of an individual who is not capable of giving informed consent, the permission of a legally authorized representative in accordance with applicable law," Guideline 4 goes on to say that "waiver of informed consent is to be regarded as uncommon and exceptional, and must in all cases be approved by an ethical review committee." The ethical committee in turn may authorize such a waiver only when a merely minimal risk comes into play and when an attempt to obtain consent would render the research unworkable.

2.3.2.2 Disclosure of "Necessary Information"

The second element of informed consent specified by the Guidelines is the disclosure of the relevant information to the potential subject. Guideline 5 provides a detailed list of all the information that should be revealed by researcher. Here I will confine myself to presenting only the most important of them.

(1) Prospective subjects should know why they are considered to be appropriate candidates for a study. They also should learn that the study is voluntary and that they are free to decline participation. A later withdrawal during the course of study also should be brought up as an option which a subject may exercise at any time and without penalty.
(2) The researcher is obliged to explain to the subject the goal of the research and the degree to which it differs from a regular medical treatment, together with the amount of time the research is expected to take.
(3) Any type of remuneration to be granted to the subject in return for participation in research ought to be made known in advance. In addition, Guideline 7 explicitly declares that certain means of recompense can never be acceptable. This refers in particular to substantial sums of money or other goods that would cause potential subjects to give consent regardless of any risks inherent in the project.
(4) After the completion of the study, subjects should be notified of its outcomes, particularly those that are pertinent to their health issues.
(5) Before the study begins, the subject must be informed of any predictable risks, pain and other difficulties the study entails. This information should also be made available to the subject's spouse or partner. Along with the risks, the benefits are to be estimated and presented as well.[36]
(6) The researcher is obliged to explain the measures to be taken in order to guarantee the subject's privacy and safeguard the confidentiality of records containing data by means of which subjects can be identified. If there are any limits to the researcher's capacity to ensure confidentiality, they should be made known to the subject along with possible consequences of breaches of confidentiality.[37]
(7) The "sponsors of the research, the institutional affiliation of the investigators, and the nature and sources of funding for the research" are to be conveyed to the subject.
(8) The subjects should know if a researcher will work during a study as an investigator only or as both the investigator and as the subject's/patient's physician.
(9) It must be clearly communicated to the subject that an ethical review committee has accepted the research protocol.

2.3.2.3 Understanding on the Part of the Subject

The next element of informed consent deals with the understanding on the part of the potential subject of the information received. This is an extremely delicate issue especially with respect to research carried out in developing countries where limited education as well as disparate cultural and social settings may deeply affect the level of subjects' comprehension. On the other hand, as has been shown, developed countries are also subject to problems related to subjects' understanding. As an illustration one may invoke the example of the assessment of certain research trials performed in Sweden, where in spite of having been informed, many participants "thought that the research procedures were part of their routine care" (Bhutta 2004, 773; see also King 2005, 140–42).[38]

2.3.2.4 Free Decision

The fourth and final element of informed consent brought up by the Guidelines centers on the decision made by the candidates to be involved in the research project. As mentioned above, this decision must remain unaffected by any type of coercion, intimidation, and undue influence. It should also be anticipated by the full and objective disclosure of relevant information, unless a partial and temporal withholding of information has been authorized by the ethical review committee. The Guidelines also qualify that as a general rule informed consent should be obtained in writing. The ethical review committee may waive this requirement only when no more than minimal risk is involved and when, for the procedures to be used, the signed consent forms are not also required beyond the research setting. A waiver could also be granted if the signed consent form seriously undermined the protection of the subject's confidentiality. Finally, in certain populations, including communal cultures, well-rooted customs deprive an investigator of direct access to potential research subjects. If this is the case, the permission of the local community authorities has to be sought prior to any other pursuits. If other native traditions exist, they also need to be respected. Nevertheless, by no means can the individual local community member's informed consent be replaced by this permission on the part of the community leaders (commentary on Guideline 4).

In summary, there is no doubt that the way in which the Guidelines address and interpret the realm of experimentation on human subjects represents a major refinement of both the Nuremberg Code and the Declaration of Helsinki. This is especially the case regarding the issue of informed consent, which is central to this study. Nevertheless, although the Guidelines respect and refer to both documents, they remain distinct and to a degree independent of them. The most significant change in relation to Helsinki VI arguably concerns the notion of informed consent. Unlike the Declaration, the Guidelines admit that the informed consent requirement, although crucial and indispensable, does not constitute by itself a sufficient protection for prospective subjects. Thus the focus on protection of subjects has been shifted from the informed consent requirement to the review responsibility of ethical committees (Howard-Jones 1982, 15; see also Perley et al. 1992, 164).

As a whole, the CIOMS/WHO Guidelines contain a considerable number of workable solutions to be applied at the level of national legislation.[39] Yet, despite their largely practical and technical character, the Guidelines are not free from ethical and philosophical premises. The main principles employed—as indicated in the introductory unit entitled "General Ethical Principles"—are those of respect for persons, beneficence/nonmaleficence, and justice. These principles were first articulated in the Belmont Report and are to be addressed in the following section.

2.4 The Belmont Report

Both the Declaration of Helsinki and the CIOMS/WHO Guidelines, because of their nature and origins, are considered fruits of international effort. Both are designed to be used in diverse research milieus throughout the world. To a degree, the same can be said of the Nuremberg Code, which, although promulgated by the U.S. judges, arose as a result of the clash between ethical standards widely accepted across nations and the Nazi ideology favoring racism and violence. Out of concern for adequate protection of human dignity, the Code took a universal perspective in its provisions.

By contrast, the document to be examined in this section is clearly an American product that developed on the tide of the public outrage at abuses in the domain of research involving human subjects in the United States after World War II. As such, *Belmont* was designed to be used in the United States. Yet, in contrast to the CIOMS/WHO Guidelines and in some respects to the Declaration of Helsinki, *Belmont* was never intended as a specific, practical, action-guiding set of principles or advisories. Rather, from the very beginning "it was meant to be a moral framework for research ethics." Thus, "*Belmont* looks to educational institutions, professional associations, government agencies, and the like to provide the particulars of research ethics" (Beauchamp 2005, 18). In what follows, I examine the Report's content through the lens of informed consent. After a brief review of *Belmont*'s historical origin I will turn to its three principles and then will outline the way they affect informed consent.

2.4.1 Belmont's Origins

For a period of 20 years following World War II, the research domain in the United States remained untouched by any major regulatory attempts. "Believing the Nazi atrocities were a world apart from the work they were doing, biomedical researchers in the U.S. between 1946 and 1966 resisted ethical and regulatory oversight as intrusive to their relationships with patients-subjects and as adequately handled at the discretion of investigators" (Vanderpool 1996, 8). This conviction of the adequacy of research practices was seriously undercut by the 1966 Henry Beecher article uncovering twenty-two instances of unethical research conduct (Beecher 1966).[40] Performed by eminent investigators, the studies in question turned a blind eye to fundamentals of research ethics. In consequence, they not only failed to protect subjects by means of informed consent, but often they also endangered their subjects' lives. Two examples provided by Beecher became particularly infamous and as such were extensively debated. The first revealed that live cancer cells were introduced into the systems of unsuspecting patients at the Jewish Chronic Disease Hospital in Brooklyn, New York. The second reported on the infection of mentally underdeveloped children with hepatitis virus at Willowbrook State School on Staten Island, New York (Vanderpool 1996, 9).[41]

In addition to the revelations provided by Beecher, appalling news on the Tuskegee Syphilis Study's victims broke in 1972 (Heller 1972). Between 1932 and 1972, four hundred black men from Alabama were involved in research without informed consent. The element of deception was actively present throughout the study as the subjects were told that they were being treated against "bad blood", the colloquial term associated locally with syphilis. In fact, "the subjects did not know that they were part of a study, did not even know what syphilis was, and did not know that they weren't being treated with available drugs" (Pence 2004, 286). One of those drugs, penicillin, became generally available as early as in 1953, and yet it was not offered to the subjects.[42]

In response to the revelation of abuses in research on human subjects and subsequent public outrage, the U.S. Congress established in 1974 the National Commission for the Protection of Human Subjects of Biomedical and Behavioral Research (hereafter the National Commission) (Brannigan and Boss 2001, 341; see also King 2005, 136), with its chief objective being "to conduct a comprehensive investigation and study to identify the basic ethical principles which should underlie the conduct of biomedical and behavioral research involving human subjects" (National Research Act 1974; quoted in Levine 2005, 126). Interestingly, the task of the Commission was to detect rather than to create the ethical basis on which the performance of research on humans could be justified. Consequently, the commissioners devoted a significant amount of time to delving into the existing guidelines on research ethics and invited a number of experts to assist in accomplishing this goal (Levine 2005, 126). The National Commission produced 17 Reports and Appendix volumes over the period 1974 to 1978 (Faden, Beauchamp, and King 1986, 215). One of its last reports and undoubtedly its most relevant was named after the site in which a closed retreat of the commissioners took place, with the aim of reflecting on the nature of basic ethical principles with whose identification they were charged by Congress. This place was Belmont House, "a conference center of the Smithsonian Institution in Elkridge, Maryland" (Jonsen 2005, 3). The fruit of the retreat and the long thought process it involved was approved by the National Commission on June 10, 1978 and published in the Federal Register on April 18, 1979 (Jonsen 2005, 5). It became known as the Belmont Report.

2.4.2 Belmont's Three Principles

The significance of the Belmont Report is anchored for the most part in its identification of the three principles meant to be a foundation for experimentation on humans. In the Report's own words, those principles "serve as a basic justification for the many particular ethical prescriptions and evaluations of human actions" (part B).[43] The three principles, that is, respect for persons, beneficence, and justice, correspond to three distinct research responsibility areas: informed consent, risk-benefit assessment, and selection of subjects.[44]

It is not clear why the Commission chose such principles and why it chose only three (cf. Lebacqz 2005, 104–108). The Document simply states in its introductory section that "three basic principles, among those *generally accepted* in *our cultural tradition*, are particularly relevant to the ethics of research involving human subjects" (emphasis added). Thus, the two correlated factors, namely "our cultural tradition" and its "general acceptability" serve as the basis for the National Commission's work.

It is debatable to what extent this somewhat intuitive foundation can be regarded as solid enough to play the role that has been ascribed to it. Additionally, a question arises as to whether the cultural tradition that *Belmont* refers to is applicable only in the American milieu or whether it applies to a broader spectrum represented by the tradition of western civilization as a whole. Since the Report begins by invoking the Nuremberg Code as prototypical of "many later codes" the latter cannot be ruled out. On the other hand, the rules provided by various codes of ethics, as *Belmont* sees them, are often "inadequate to cover complex situations; at times they come into conflict and they are difficult to interpret or apply." Therefore, "broader ethical principles," that is, those proposed in the Document, "provide a basis on which specific rules may be formulated, criticized and interpreted" (introductory part).

Although it remains unclear which codes of ethics the authors of the Report have in mind, they seem to have conceived their job in terms of providing an alternative. Consequently, they refer to American cultural tradition in a way that presupposes something different from and complementary to the existing trend. Despite that, the Report preserves a certain distance from its principles and acknowledges that principles other than these three may turn out to be pertinent to the area of research on humans. Nonetheless, *Belmont* sees its principles as "comprehensive ... and stated at a level of generalization that should assist scientists, subjects, reviewers and interested citizens to understand the ethical issues inherent in research involving human subjects." Even though the principles cannot always be applied so as to resolve beyond doubt any single dilemma, "the objective is to provide an analytical framework that will guide the resolution of ethical problems arising from research involving human subjects" (introductory part).

Despite being introduced as an alternative to the rules provided by existing codes of ethics and despite their incontestable attractiveness, *Belmont*'s principles are far from perfect. In reality, they struggle with problems similar to those found by the Report in other ethical codes, which is to say that they conflict with each other and at times are difficult to interpret or apply. The crucial difficulty identified by Robert Veatch is that *Belmont* does not address "the question of whether for a research protocol to be morally justified, all the principles must be satisfied or whether, when they conflict, the principles taken together must merely be satisfied 'on balance.'"

This issue is of the utmost importance because "without addressing it, the *Belmont* principles are virtually meaningless" (Veatch 2005, 190). Indeed, it permeates deeply into the nature of principles: are they merely prima facie duties, which implies their equal status and the potential for one principle to be overridden by either one or both of the two that remain, depending on the concrete situation? Or perhaps a certain hierarchy exists among the principles so that, for instance, the principle of respect for persons should be regularly given precedence over the other two? The former interpretation is usually regarded as more plausible by commentators, since, as stated by Robert Levine, "there was no intent on the part of the National Commission to establish a lexical ordering" (Levine 2005, 127) among the principles. Nonetheless, Albert Jonsen believes that a careful reading of the Report leads to the conclusion that "even without a specific articulation of rank, it is clear that the first of these principles dominates the other two" (Jonsen 2005, 8). In my opinion, both statements appear defensible in light of *Belmont*. This leaves a great deal of room for interpretation that may vary significantly depending on the philosophical predilection and intellectual preferences of interpreters.

No doubt, it is primarily thanks to their broad scope and flexibility that the principles play such a significant role in contemporary bioethics. It also goes beyond saying that *Belmont* inspired multiple philosophical applications and developments. In fact, the principles have had a considerable impact outside the academic setting as well. It is hard to overestimate the influence that the work of the National Commission, of which *Belmont* is a significant part, exerted on policies regarding the protection of human subjects. The examples thereof can be "seen in the education and training programs that include questions on *Belmont* required by National Institutes of Health (NIH)-funded investigators at U.S. institutions, in the assurances negotiated between universities and the federal government that often include *Belmont* as an appendix, and in the required readings in university courses on biomedical ethics" (Shapiro and Meslin 2005, 56–57). The inclusion of *Belmont*'s principles in the CIOMS/WHO Guidelines shows that the principles' influence goes beyond the U.S. research milieu.

2.4.2.1 The Principle of Respect for Persons

The first principle presented in *Belmont* is the principle of respect for persons, which "incorporates at least two ethical convictions: first, that individuals should be treated as autonomous agents, and second, that persons with diminished autonomy are entitled to protection" (part B, n. 1). The Report clearly speaks of respect for persons in terms of safeguarding their autonomy. Apparently for this reason, there is a tendency to equate the principle of respect for persons presented by *Belmont* with the principle of autonomy or respect for autonomy, especially as captured by Beauchamp and Childress in their *Principles of Biomedical Ethics* (Beauchamp and Childress 2001). This reckoning, however, as has been stated, stands at odds with the National Commission's intentions (Levine 2005, 128). It is known that "there are attributes of persons other than autonomy that create a moral obligation to treat them with respect" (Levine 2005, 128).[45] According to the National Commission's view "*all* people, *all* human beings are deserving of respect" (Lebacqz 2005, 100). In other words, no matter to what extent a human being enjoys autonomy, that being merits respect.

It is noteworthy that *Belmont* does not distinguish between human beings and persons—it uses both terms interchangeably. This is not unimportant, since some scholars define persons wholly through the prism of their capacity for autonomous choices (Lebacqz 2005, 100). The language employed by *Belmont* as well as the views of its authors leave no doubt that human beings do not have to be in possession of a certain amount of autonomy in order to be considered persons. Consequently, no matter whether one refers to man as person or as human being or in any other way, the one thus referred to deserves respect.

Nonetheless, the placement of the protection of "those with diminished autonomy" in the section dedicated to the principle of respect for persons has been criticized by certain authors and even considered to be flawed (see e.g. Veatch 2005, 187). Beauchamp and Childress, for instance, locate it under the principle of nonmaleficence (see Beauchamp and Childress 2001, 152–58). Although it is plausible to maintain that recognizing the need for safeguarding subjects with lessened autonomy within the boundaries of the principle of respect for persons raises questions, it is no less plausible to note that by doing so *Belmont* points precisely to the fact that there is more involved in this principle than merely human autonomy. I believe that no fewer than two additional factors exist to support this line of reasoning.

First, as Commissioner Karen Lebacqz has pointed out, despite advancing the evident link between respect for persons and autonomy, the National Commission clearly stated that "the principle of respect for persons incorporates *at least* basic convictions about autonomy; however, it left room of additional understandings of the importance of respect for persons that might extend beyond respect for autonomy" (Lebacqz 2005, 100). In other words, it is, among other things, the wording "at least" employed by the Commission that does not allow for elimination or marginalization of various components of the principle of respect for persons other than autonomy. Unfortunately, the Commission never articulated what components apart from autonomy are to be involved in this principle (Veatch 2005, 187).

Second, as highlighted by Larry Churchill, *Belmont*'s interpretation of autonomy can be deemed weak and open to question (Churchill 2005, 111). The Report declares, for instance, that "to respect autonomy is *to give weight* to autonomous persons' *considered opinions and choices* while refraining from obstructing their actions unless they are clearly detrimental to others" (part B, n. 1, emphasis added). According to Churchill, the "*Belmont*'s stress on 'giving weight' is anemic in comparison to the usual understanding of respect for autonomy as recognizing that individuals have final authority for decisions or that they are free and self-determining moral agents" (Churchill 2005, 111). Thus the Report seems to reallocate the moral agency away from subjects towards those who experiment upon them. Moreover, the stress on "considered opinions and choices" becomes problematic insofar as one realizes that a subject's ill-considered or unconsidered opinions should perhaps also be recognized and respected. This seems obvious as regards research, when subjects "should be free to decline participation for any and all reasons—considered, well-considered, ill considered, or completely unconsidered" (Churchill 2005, 112).

Churchill maintains that the "considered-opinions" problem reveals that the notion of autonomy derived from the therapeutic context was used by the National Commission, which in turn leads to bewilderment and confusion when it is applied to the research domain. If this is the case, then the boundaries between the practice of accepted therapy and research, carefully drafted in the first part of *Belmont*, might lose their significance and clarity. The Report states that "it is important to distinguish between biomedical and behavioral research on the one hand, and the practice of accepted therapy on the other, in order to know what activities ought to undergo review for the protection of human subjects of research. The distinction between research and practice is blurred partly because both often occur together (as in research designed to evaluate a therapy) and partly because notable departures from standard practice are often called 'experimental' when the terms 'experimental' and 'research' are not carefully defined" (part A).

If Churchill is right, the distinction between research and practice is also blurred by the language used by *Belmont* in the section devoted to respect for persons. Thus, the Report does not manage to escape from the trap which it sees elsewhere and which it warns against at the outset. Nevertheless, regardless of whether or not or to what extent Churchill's judgment regarding "considered-options" problem is justified, applying a weak notion of autonomy is not necessarily a drawback. Indeed, *Belmont*'s approach shows that such a notion can be legitimately used as a meaningful tool in referring to the respect for persons. Autonomy, as the Report depicts it, does not exhaust the total significance of being a person and therefore it contributes to and subjects itself to rather than replacing respect for persons. For this reason, instead of calling it weak, I propose that the notion of autonomy introduced in *Belmont* be labeled limited. Its limitations may be observed particularly in relation to persons who for different reasons cannot enjoy its fullness. By contrast, any strong, close-to-absolute notion of autonomy has the potential for undermining rather than bolstering the respect to which every person, even those who are not fully autonomous, is entitled. This might be, I believe, the reason why the Commissioners chose neither to advance a strong notion of autonomy nor to favor it as the ultimate yardstick by which to measure the respect for persons.[46]

Whether seen as strong, weak, or limited, the notion of autonomy has not itself been defined by the Report. Since the Commissioners represented different intellectual bents and outlooks on life, they did not "even attempt to buttress the Belmont principles with a theory" (Beauchamp 2005, 19). Although not unreasonable, this reluctance generates a number of problems. For instance, the Report uses the words "autonomy" and "self-determination" interchangeably, which gives an impression that the two are regarded synonyms. In fact, a substantial difference exists between, for example, the notion of autonomy espoused by John Stuart Mill and that of Immanuel Kant. The former corresponds to a libertarian view according to which individuals should be allowed to do freely whatever they wish for themselves, while the latter entails the element of implicit universalization and thus puts certain limits on unfettered freedom of choice. At the level of language, Mill's notion of autonomy parallels what is now identified as self-determination, whereas Kant's concept can be more properly called "autonomy" (see Lebacqz 2005, 101).

As mentioned above, the principle of respect for persons finds its practical application in securing informed consent. *Belmont* sees consent as a process involving three elements: information, comprehension, and voluntariness. Regarding the information, different standards exist with respect to the quality and quantity of knowledge to be provided to subjects. *Belmont* embraces a skeptical position as to the "commonly provided information standard" (what kind of data is conveyed to subjects by the majority of practitioners in a given field) as well as to "the reasonable-person standard" (what a reasonable person would want to know in order to make a choice regarding proposed medical procedures). The standard favored by the Report is that of "the reasonable volunteer," according to which "the extent and nature of information should be such that persons, knowing that the procedure is neither necessary for their care nor perhaps fully understood, can decide whether they wish to participate in the furthering of knowledge" (part C, n. 1).

When it comes to comprehension, the second element required in the process of informed consent, the Report states that "the manner and context in which information is conveyed is as important as the information itself" (part C, n. 1). The researchers are charged with ascertaining that the information has been adjusted to the subjects' capabilities and conveyed in such a way as to be understood. In situations where subjects' ability to comprehend is diminished, a proxy should be involved with the research in order to decide on behalf of a subject which procedures safeguard the subject's best interest.

The last element of the informed-consent process, voluntariness, "requires conditions free of coercion and undue influence." It is noteworthy that *Belmont* attempts to define those two terms. It states that "coercion occurs when an overt threat of harm is intentionally presented by one person to another in order to obtain compliance. Undue influence, by contrast, occurs through an offer of an excessive, unwarranted, inappropriate or improper reward or other overture in order to obtain compliance. Also, inducements that would ordinarily be acceptable may become undue influences if the subject is especially vulnerable" (part C, n. 1). These definitions make up a lucid and functional means by which to guarantee the subjects' voluntariness. Yet, as *Belmont* admits, certain circumstances still entail dilemmas that are very difficult to resolve. An example offered refers to engaging prisoners as research subjects. "On the one hand, it would seem that the principle of respect for persons requires that prisoners not be deprived of the opportunity to volunteer for research. On the other hand, under prison conditions they may be subtly coerced or unduly influenced to engage in research activities for which they would not otherwise volunteer. Respect for persons would then dictate that prisoners be protected" (part B, n. 1).

I believe this example sheds valuable light on Churchill's critique of the way in which the phrase "give weight" has been used by *Belmont* in relation to autonomy. To illustrate this in a simple way, even if prisoners want to participate in research, their will does not necessarily constitute the only and absolute factor to be taken into account. No doubt, in many daily-life situations outside the prison wall, one may also be "subtly coerced or unduly influenced." The autonomy of every person, including possible research participants, may indeed be affected by various components, often not only unknown to investigators but also not even realized by the potential research subjects—a phenomenon that should be subject to thorough scrutiny during the informed-consent process.

For this reason, investigators are better able to protect and promote their subjects' autonomy in two ways. First, they can give weight to "'autonomous persons' considered opinions" rather than treating those opinions as an absolute indicator of what should be done. Second, they can examine potential and often barely detectable factors threatening real freedom of choice. Thus, the informed-consent process based on understanding autonomy as limited more fully respects persons: it helps subjects become conscious of the possible constraints of their autonomous choices and in this manner truly serves their own interest. By no means does it disregard individuals' decisions. Rather it supports them by enabling more informed and conscious choices.

2.4.2.2 The Principle of Beneficence

The second principle selected by *Belmont* is that of beneficence. Understood as an obligation, this principle plays a crucial role in justifying research on humans. It comprises two fundamental maxims: (1) "do not harm" and (2) "maximize possible benefits and minimize possible harms" (part B, n. 2). As has been stated, this phrasing "encompasses both doing good and avoiding harm" which "is certain to lead to confusion" (Veatch 2005, 186). In order to avoid confusion, numerous scholars proposed the recognition of a fourth principle, namely that of nonmaleficence, as distinct from beneficence (see Beauchamp and Childress 2001, 114–16; see also e.g. Ross 1988, 21), but this is not *Belmont*'s position. In addition, a separate issue surfaces regarding whether it is doing good or avoiding harm that should be given priority in ethical assessments of research projects. Some moral thinkers, like David Ross, favor not doing harm, considering it more relevant and binding than doing good (Ross 1988, 21; as referenced in Veatch 2005, 186). Others, such as William Frankena, perceive the two as "merely two dimensions of the same variable, one positive and the other negative" (Frankena 1973, 47). Accordingly, "benefits and harms can be combined on the same scale" (Veatch 2005, 186). This position reflects, at least to a substantial degree, *Belmont*'s approach to dealing with avoiding harm and doing good.

After recognizing the crucial role that the principle of beneficence has been playing in medicine since Hippocrates, the Report invokes Claude Bernard, who applied it to research, with an emphasis put on the fact that the injuries done to the subject should never be offset against nor justified by the expected benefits resulting from research. Indeed, injuries should never happen. In a sense, *Belmont* takes up Bernard's claim, develops it and clarifies it. From the outset, the Report says that "even avoiding harm requires learning what is harmful; and, in the process of obtaining this information, persons may be exposed to risk of harm." Likewise, "learning what will … benefit [the subjects], may require exposing persons to risk." The actual dilemma then can be framed as follows: when is it "justifiable to seek certain benefits despite the risks involved" and when should the benefits be forgone on account of accompanying risks? (part B, n. 2).

In order to answer these questions one needs not only to know what exactly presents risks and what promises benefits but also how to balance risks and benefits so as to determine their ratio. This risk-benefits assessment makes up the practical application of the principle of beneficence and as such parallels the informed consent that is derived from the principle of respect for persons. The assessment must demonstrate that risk and benefits exist "in a favorable ratio," which presumably requires that the latter outweigh the former. Unfortunately, *Belmont* does not offer much help in specifying what is meant by "favorable ratio." In fact, a number of further difficulties emerge as regards the measures to be taken in order to determine such a ratio. *Belmont* admits, for instance, that precise judgments of this type are almost unattainable. Yet, the Report also maintains that "the idea of systematic, nonarbitrary analysis of risks and benefits should be emulated insofar as possible." What is more, an explicit method needs to be employed in the process of risk assessment, "especially where there is no alternative to the use of such vague categories as small or slight risk." Additional steps need to be taken to ascertain "whether an investigator's estimates of the probability of harm or benefits are reasonable, as judged by known facts or other available studies" (part C, n. 2).

2.4.2.3 The Principle of Justice

In addition to respect for persons and beneficence, the Report presents a third principle—justice. Fundamentally, as seen by *Belmont*, the principle of justice addresses two problems: bearing the burdens of research and distributing its benefits. Thus, "an injustice occurs when some benefit to which a person is entitled is denied without good reason or when some burden is imposed unduly" (part B, n. 3). Even with this explanation, it still remains unclear who is entitled to what and on what basis. A useful tool in the quest for clarification relates to another way of understanding the principle of justice provided by the document. In short, the principle comes down to saying that "equals should be treated equally." In order to determine who is equal and who is unequal the Report refers to "several widely accepted formulations of just ways to distribute burdens and benefits . . . These formulations are: (1) to each person an equal share, (2) to each person according to individual need, (3) to each person according to individual effort, (4) to each person according to societal contribution, and (5) to each person according to merit" (part B, n. 3).

In addition to, and closely related to, the issue of fair distribution, *Belmont* calls for the just selection of research subjects. As history has shown, the burden of being involved in research as subjects has frequently fallen to underprivileged, poverty-stricken, or institutionalized persons, whereas the research benefits flowed to the well-off. Therefore, "whenever research supported by *public funds* leads to the development of therapeutic devices and procedures, justice demands both that these not provide advantages only to those who can afford them and that such research should not unduly involve persons from groups unlikely to be among the beneficiaries of subsequent applications of the research" (part B, n. 3, emphasis added).

Inexplicably, the Report restricts this provision to publicly funded research and thus renders it difficult to deduce what standard applies to research supported by private sources. Does such research escape the requirement of fair selection of subjects? The answer sought in the spirit of *Belmont* seems to be negative, but this does not change the fact that by referencing public funds, the Document exposes itself to the risk of misinterpretation or confusion. This risk becomes even greater when one realizes that "the boundaries between publicly and privately funded research are by no means clear or complete" (Sherwin 2005, 159).

Fair selection of research subjects, as seen by the Report, is both an application of and a consequence of the principle of justice. This parallels informed consent springing from the principle of respect for persons as well as the risk-benefits assessment derivative of the principle of beneficence. In a word, justice understood in relation to the selection of subjects can be examined on two levels: individual and social. The first level points to the way in which individual subjects are chosen, while the second refers to various classes of subjects, their availability, vulnerability, and suitability to be enrolled in research. Nondiscriminatory application of social justice requires that "there is an order of preference in the selection of classes of subjects (e.g., adults before children) and that some classes of potential subjects (e.g. the institutionalized, mentally infirm or prisoners) may be involved as research subjects, if at all, only on certain conditions." *Belmont* emphasizes that no matter to which level of justice one refers, partiality or a range of "social, racial, sexual and cultural biases" might pose a serious threat to justice. Thus, "even if individual researchers are treating their research subjects fairly, and even if IRBs are taking care to assure that subjects are selected fairly within a particular institution, unjust social patterns may nevertheless appear in the overall distribution of the burdens and benefits to research" (part C, n. 3).

By this and by the other standards mentioned above, *Belmont* handles justice-related issues in a serious and responsible manner. Still, the strict link between justice and selection of subjects it proposes, although useful, narrows the field of vision almost entirely to distributive justice. Other types of justice occupying well-defined positions in bioethics literature, such as, for instance, compensatory or procedural justice, are overlooked in the document. Compensatory justice requires redress for harm or wrong done to research subjects. Procedural justice draws attention to the fact that radical exclusion of vulnerable groups from research amounts under some circumstances to the denial of the benefits flowing from research. Thus, a surfeit of protection is likely to generate a different form of injustice. In other words, an effective implementation of the principle of justice is gravely challenged not only by *Belmont*'s well-described overrepresentation of certain groups in research trials but also by the corresponding under-representation of other groups. In order to counteract any possible form of discrimination, vulnerable subjects such as women and minorities must be protected in such a way so as not to be deprived of the chance to participate in research and in the benefits that flow from it. To achieve this goal, appropriate procedures should be established with respect to decision-making policies at both the institutional and the national level (see King 2005, 142–45).

These inadequacies notwithstanding, the way in which the Report copes with the principle of justice is worthy of appreciation. As has been acknowledged, "*Belmont* was the first public medical ethics document to explicitly recognize that justice is morally relevant in human subjects research" (Veatch 2005, 188). Given the time of its origination, the Report presented the principle of justice fairly and thus anticipated the place it has occupied in bioethics over recent decades.

2.4.3 Informed Consent and the Three Principles

The question of how informed consent relates to the *Belmont* principles will be considered in this subsection. The major portion of this issue has already been covered when discussing the principle of respect for persons. As has been shown, the Report introduces a strong and direct link between informed consent and the principle of respect for persons. By contrast, in contemporary bioethics literature a strong bond between the principle of autonomy and informed consent prevails. As stated above, *Belmont*'s principle of respect for persons, even if defined by means of autonomy, does not preclude other components. Consequently, it does not conform to a strong notion of autonomy. It rather holds to its limited version of autonomy as combined with other concepts constitutive of being a person. Therefore, autonomy should not be regarded as the only or even the most important aspect to be taken into consideration when securing informed consent. To put it another way, informed consent is not ensured by the mere fact that the potential subject's autonomy has been respected.

Furthermore, *Belmont* recognizes the connection between informed consent and the principle of beneficence that occurs basically at the procedural level. By stating that "relevant risks and benefits must be thoroughly arrayed in documents and procedures used in the informed consent process" (part C, n. 2), the Report depicts the risk-benefits assessment as a vital element of the informed consent required for research validation. On the other hand, both risks and benefits are to be made known to the subjects in a comprehensible way which, in turn, touches on the principle of respect for persons.

When it comes to the principle of justice, pinpointing its relation to informed consent presents a greater difficulty. *Belmont* does not direct that potential subjects should be informed as to why they are regarded as suitable candidates to participate in research. If this were required, as it is, for instance, in the CIOMS Guidelines, a solid link to the issue of selection of subjects would be set up. Although not explicit and not easily identifiable in *Belmont*, the connection can be demonstrated by examining its treatment of the topic of vulnerable subjects. Generally, the Report speaks of vulnerable subjects at least in two senses. The first raises concerns as to the possibility of exploitation, due, for instance, to compromised economic or social position. Yet subjects who belong to these groups are capable of giving a valid informed consent. The second sense in which one may refer to vulnerability relates to situations in which obtaining meaningful consent is hampered or even impossible. "To the extent that vulnerability is linked with questions of diminished capacity to consent, inability to protect oneself through the consent process, or inability to give informed consent, concern for selection of subjects overlaps with the principle of respect for persons" (King 2005, 139). At the same time, however, as noted above, concern for selection of subjects immediately applies to the principle of justice. Hence, to show the link between informed consent and the principle of justice, a syllogism can be applied. If vulnerability plays a crucial role in the selection of subjects and if the lack of informed consent causes vulnerability (at least in one sense as seen in *Belmont*), then the conclusion is that the lack of informed consent strictly relates to the principle of justice.

In summary, the connection between informed consent and the principle of respect for persons is direct and primary, while it is indirect and secondary with respect to other principles. This also seems to be the case regarding proxy consent. The Report affirms that persons not capable of self-determination deserve protection (see part B, n. 3). Thus, although not mentioned explicitly in the text, proxy consent as a conceivable element of such protection falls under the principle of respect for persons. Much the same can be said of the principle of beneficence, since the procedural requirement invoked above clearly applies to informed consent and thus also to its proxy form. Likewise, as far as the principle of justice is concerned, the way it relates to informed consent forms the basis for proxy consent. The inability to give informed consent of the required quality usually calls for a surrogate, insofar as inclusion of vulnerable subjects into the research projects is not precluded on other grounds.

2.5 Conclusion

Informed consent as related to research involving human subjects was examined in the previous chapter through the lens of its historical origins, including its early appearance in Catholic tradition. This chapter has built on that foundation by analysis of the human-rights documents of central importance to this matter. Obviously, these documents "reflect the concerns and perspectives of the time in which they were written and of the individuals by whom they were written" (Levine 1996, 250). In effect, their content hinges to a great degree on the purposes for which they originated. Accordingly, each document presents the theme of research involving humans as subjects in a different way.

In contrast to its successors, the Nuremberg Code "was not the outcome of an attempt to frame new principles of medical ethics, but rather a formulation, in the course of a trial for war crimes, of criteria said to be widely accepted by the medical profession, against which the acts of certain physicians carried out on prisoners might be judged" (Howard-Jones 1982, 1436). As such, *Nuremberg* manifests a fairly narrow scope when considered against what currently pertains to research ethics. It neither directly addresses research on vulnerable populations nor provides for proxy consent. What is more, it also does not mandate independent ethical review boards. The main instrument by which to protect the subject's safety has been identified with a strong and rather restrictive notion of informed consent articulated immediately at the outset of the Code. Thus perceived, the concept of informed consent tends to restrict many research activities, even to the extent of their complete exclusion. Interestingly, even though subjects preserve the right to refuse participation and to withdraw from research, everyone involved in the research project bears responsibility for determining the quality of informed consent. Subsequently, it is investigators and research participants rather than the institutions for which they work that are ultimately held accountable.

Despite its generally rigorous character, the Nuremberg Code postulates fairly surprisingly that although "no experiment should be conducted where there is an a priori reason to believe that death or disabling injury will occur," exceptions to this rule might perhaps be allowed if "the experimental physicians also serve as subjects" (art. 5). As Howard-Jones pointed out in his 1982 article, "no subsequent or previous bioethical code has included such a waiver" (Howard-Jones 1982, 1436). As far as I have been able to ascertain, this conclusion has remained true up to the present.

While the Nuremberg Code speaks literally of "voluntary consent" and sees "informed" as one of its basic characteristics, the Declaration of Helsinki may be the first human rights document to utilize the phrase "informed consent" (Levine 1996, 242). This change in terminology refers to the fact that making a free decision necessitates obtaining the relevant information prior to the decision. At the same time, by not favoring the term "voluntary consent," the Declaration distances itself from the heritage of World Word II in which the memory of countless subjects involuntarily used in the conduct of appalling experiments occupied a central position.

In contrast to the Nuremberg Code, *Helsinki* distinguishes between therapeutic and nontherapeutic research and also provides for proxy consent. In addition, it creates the possibility of a legitimate consent waiver. Aside from the potential danger to subjects' safety incurred by the practical application of the first two innovations, the third change clearly relaxes the consent requirement. The allowance for a consent waiver is left to the sole discretion of the ethics committees and is not protected by a supplementary set of reliable safety measures. Along with several other provisions which were examined in the earlier section dedicated to the Declaration's content, this demonstrates an initial devaluation of the notion of informed consent, so robustly introduced by the Nuremberg Code.

One of the thorniest ethical questions tied in with the human-rights documents regarding research involving humans as subjects is that of ethical universalism and pluralism. In short, "ethical universalists believe there is a universal set of ethical principles that are applicable to all human beings regardless of their situations in particular cultures. The task of the moral philosopher, then, is to discover those universal principles that apply in all times and in all places. Ethical pluralists, by contrast, recognize that all ethical principles are developed in the course of discussions held within particular cultures and that these discussions necessarily reflect the unique histories and other circumstance of particular cultures. In this view, ethical principles are invented rather than discovered" (Levine 1996, 237).

In the light of this clarification, it becomes obvious that the authors of the Nuremberg Code, by recalling that universal agreement exists as to the basic moral principles, employed reasoning based on natural law with respect to informed consent and the entire field of experimentation on human subjects. Much the same can be said of the Declaration of Helsinki, the approach of which—on the one hand highly influenced by *Nuremberg*, on the other vulnerable to the deflation process—preserves the universal perspective. Interestingly, the CIOMS/WHO Guidelines, although meant to provide concrete ways of applying *Helsinki*, appear to some to move in a different direction. No doubt, it is precisely in regard to the Guidelines that the dispute between universalists and pluralists finds its sharpest expression. This is clearly illustrated in the conflict over using placebos in trials carried out by researchers from developed countries but hosted by developing countries, and the approval of procedures that would never be agreed to in developed countries.

Levine summarizes this by saying that "like *Nuremberg* and *Helsinki*, the CIOMS Guidelines also aspire to worldwide applicability. However, unlike *Nuremberg* and *Helsinki*, they do not assert the universal validity of a set of rules that either prescribe or proscribe specific behavior. Rather they recognize the legitimacy of a limited amount of ethical pluralism" (Levine 1996, 238). The Guidelines also recognize the fact that informed consent is a process rather than a single event and insist on the need for its periodic renewal even if there are no modifications regarding the research conduct or its goals. On the other hand, however, the consent as perceived by the Guidelines is far less powerful in comparison with that of the Nuremberg Code and the Declaration of Helsinki. In some measure, the document itself directly diminishes the meaning of consent, as can be illustrated by recalling its proxy form. As mentioned earlier, the low-risk standard rather than minimal-risk standard applied to proxy consent show a degree of lenience which has generated significant controversy. Perhaps for this reason, the Guidelines do not consider informed consent a sufficient means by which to safeguard research subjects. The emphasis clearly moves toward the role ethical review committees are supposed to play in ascertaining a satisfactory implementation of ethical standards, a stance that clearly distinguishes the Guidelines from its predecessors. Thus, by their distrust in the sufficiency of informed consent, the CIOMS Guidelines effectively continue the devaluation of informed consent, a trend first clearly observable in the Declaration of Helsinki.

Despite the unquestionable differences existing between the Nuremberg Code and *Helsinki* on the one hand and the Guidelines on the other, the former are invoked with respect and appreciation by the latter in one of its introductory chapters. Likewise, the Belmont Report, although not explicitly mentioned by the Guidelines, exerted its influence on them via the three principles. Still, the extent to which the two documents are in actual fact connected with each other remains uncertain. Similarly, the rather vague relationship between the three principles employed in the opening section of the Guidelines and their subsequent content calls for further investigation.

Unlike *Belmont*, the CIOMS Guidelines aim at concrete applications of ethical research solutions in the international context. In this respect they also differ from both *Nuremberg* and *Helsinki*. Still, at least three further dissimilarities exist between those two documents and the Belmont Report. First, *Belmont*, in accord with the CIOMS Guidelines, places the responsibility for research on institutions rather than on investigators. Second, unlike the Declaration of Helsinki, it offers no requirement for experimentation on animals as preceding and justifying that on human subjects. Third, in harmony with the Nuremberg Code the Report employs a limited rather than a close-to-absolute notion of autonomy. Still, unlike *Nuremberg*, it provides for proxy consent (Winslade and Krause 1998, 150). Fourth, as opposed to the CIOMS Guidelines and the Helsinki Declaration, both *Belmont* and *Nuremberg* have never been revised and updated.

In the case of the Nuremberg Code, this lack of subsequent revision corresponds not only to its historic origins but also to the role the Code has played over decades in the field of research involving human subjects. Despite having been branded as "a polar star of ethics in biomedical research" (Herranz 1998, 127) by some and being spurned as obsolete and out of sync with the current research reality by others, the Code still seems to have something significant to say for those engaged in present-day research involving human subjects. As a matter of fact, the heritage of *Nuremberg* has never been exhausted. This observation, initially made by Herranz, is especially relevant to two issues.

First, at the end of its opening clause *Nuremberg* imposes an ethical (not legal or punitive or civic) duty for "ascertaining the quality of consent" on all the members of the research team. This obligation has been defined as "personal" and as such cannot be delegated without sanctions. In line with Herranz's comment, "consenting or non-consenting is thus not exclusively up to the subject but also involves an ethical ruling by the researcher" (Herranz 1998, 136).

Second, article 9 of the Code declares that a subject undergoing an experiment "should be at liberty to bring the experiment to an end if he has reached the physical or mental state where continuation of the experiment seems to him to be impossible." Remarkably, this provision is conditional. In contrast to the subsequent codes allowing for an absolute right to withdraw from the research enterprise at any time and even if only by virtue of the subject's caprice or whim, it sees subjects as moral beings responsible for their actions. From this perspective, informed consent is binding not only for investigators and institutions conducting research but also for subjects themselves.

As Herranz points out, none of the subsequent human-rights documents dealing with human experimentation sufficiently take into account these two provisions of the Nuremberg Code. In fact, they seem to have been either forgotten or deliberately omitted. This is clearly unfortunate and not easy to understand because the two Nuremberg postulates touch on the very core of research ethics.

The first postulate stands at odds with the *Helsinki*'s minimalist comprehension of the moral role played by investigators and the entire research team. In fact, it is the job of ethical committees rather than of investigators to ascertain that the subjects' consent meets all the ethical criteria. As stated above, CIOMS Guidelines go further, even to the point of challenging the adequacy of informed consent, and thus they give even more authority to ethics committees.

Furthermore, it is not only the investigator's moral role that has been diminished by *Helsinki* and subsequent ethics codes. In fact, the subject's moral role has also been perceived in a minimalist way which opposes the second *Nuremberg* postulate. With the purpose of facilitating the process of securing informed consent, the subject "is given a permanently open door by which to leave the experiment when he sees fit without having to give any reasons." By contrast, *Nuremberg* takes the subject's dedication to the research project far more seriously. As Herranz put it, "for Helsinki, unlike Nuremberg, the consent given by the subject is not binding in any way: it is always a provisory decision, if not arbitrary or capricious. The subject's dedication to the project can be retracted at any moment." Conversely, "consent as defined in Nuremberg requires the addition of indispensable information about the consequences resulting from the subject's decision to withdraw from the experiment. This information makes the consent clearer, more voluntary and rational than the consent of the Helsinki type, which fosters a less responsible kind of behavior" (Herranz 1998, 137).

Thus, two distinct notions of informed consent emerge: the Nuremberg type and the Helsinki type. While the former has been abandoned, the latter continues to be significant in contemporary bioethics. Yet, two additional characteristics of particular importance for this study, stemming from the failure to apply the aforementioned Nuremberg postulates, can be delineated.

First, there is no attempt in the human-rights documents analyzed in this chapter to reconcile the subject's right to quit the experiment with the moral responsibility he or she bears for giving informed consent before enrolling in a research project. Second, there is no word on the moral integrity of the research subjects and of all the research team members. In other words, what is missing is the lack of recognition of the need for moral formation of future investigators. Ethical formation signifies much more than mere education aiming at technical skills combined with an acquaintance with legal requirements and medical procedures. As the later parts of this study will show, moral integrity proves to be a critical factor for the successful and subject-safe conduct of experiments, especially when referring to the investigators in charge. Unfortunately, the ethical codes and guidelines fail to recognize this. In fact, the dominant logic on which they operate both implies and supports an incorrect conviction that it is the procedures utilized in research rather than the moral quality of those who employ them that play a pivotal role in the attempt to render research on human beings ethical. This is most definitely an approach with which I cannot concur. By the analysis which constitutes the content of the later chapters of this work, I intend—among other tasks—to explain why.

Chapter 3
The Major Current Interpretations of the Principle of Free and Informed Consent

While the first chapter of this study focused primarily on the historical origins of informed consent and its second on the articulation of this concept in human-rights documents, the purpose of this chapter is to trace the issue of informed consent throughout the Magisterial teaching of the Roman Catholic Church (first section) and its treatment in relevant philosophical and theological literature (second section). This will provide grounds for comprehensive analyses of proxy consent. Space limitations require that the second section include only the most representative and recognized authors. Thus, the virtue ethics applied to the medical realm by Edmund Pellegrino and David Thomasma will be contrasted with the four-principle approach characteristic of Tom Beauchamp and James Childress, enriched by psychological insights provided by Ruth Faden. The account of Paul Ramsey, rooted in the biblical notion of covenant and classic for the topic, will open the section, while the analyses of Germain Grisez, grounded in Catholic theology and emphasizing the weight of conscience, will close it. Furthermore, the contributions of authors such as Benedict Ashley, Kevin O'Rourke, Robert Veatch, Jessica Berg, Paul Appelbaum, Charles Lidz, Lisa Parker, and Robert Levine will be recognized as appropriate throughout this section.

As clearly acknowledged in current bioethics literature, informed consent, far from being an absolute moral requirement, allows for ethically and legally valid exceptions under certain circumstances. In the third section these exceptions will be identified and examined. The fourth section, in turn, will offer a concise introduction to the issue of proxy consent, principally by clarifying terminological ambiguities and recognizing the significance of the distinction between therapeutic and nontherapeutic research.

3.1 Relevant Magisterial Teaching

A number of documents of the Magisterium deal with the issue of informed consent. These documents have varying levels of importance. The most important, of course, are those of the Pope, e.g., addresses by Pius XII and John Paul II, and documents

G. Mazur, O.P., *Informed Consent, Proxy Consent, and Catholic Bioethics*, Philosophy and Medicine 112, DOI 10.1007/978-94-007-2196-8_3,

issued by Pontifical Congregations and Councils, e.g., those of the Congregation for the Doctrine of the Faith. These highly important documents, however, while briefly addressing the issue of informed consent, do not do so in a systematic way. Thus I present here two Magisterial documents that arguably carry the greatest significance in this respect, namely the *Charter for Health Care Workers* and *Ethical and Religious Directives for Catholic Health Care Services.* Both of these documents draw abundantly from a rich repertory of Church teaching, in particular papal addresses. Other Magisterial documents that refer to informed consent do so explicitly in the context of organ transplantation (informed consent required from both donor and recipient; see John Paul II 2000), sterilization (decisive critique and rejection of forced sterilizations carried out without genuine informed consent; see Pontifical Council for the Family 1998), and marital love (communion of life and love between husband and wife commences with their informed and free consent; see John Paul II 1996).

Given their various origins, methodologies, and purposes, it is not easy to contrast the Magisterial documents to be analyzed in the present section with the human-rights documents presented in the first chapter. Nor is this contrast the aim of this study. Nevertheless, although it is not my intention to engage in an exhaustive comparison, I am convinced that a juxtaposition of certain aspects, even if not comprehensive, is of great value, as it will offer a better understanding of both differences and similarities between the two groups of documents.

3.1.1 Charter for Health Care Workers

Published by the Pontifical Council for Pastoral Assistance to Health Care Workers in 1995, the *Charter for Health Care Workers* has been described by Michael Prieur as "a 'gold mine' of official papal teachings, organized in a fresh and inspiring way" (Prieur 1996, 15). Moreover, it refers creatively and at some length to such Magisterial documents as the *Declaration on Procured Abortion* (1974), the *Declaration on Euthanasia* (1980) and the *Instruction on Respect for Human Life in Its Origin and on the Dignity of Procreation*, known as *Donum Vitae* (1987). "The Charter breaks new ground by introducing some new concepts and principles related to moral theology" and by doing so it "reflects twentieth century developments" in the field. Among the concepts explicitly bearing on informed consent, the principles of therapeutic trust, responsibility for health care, and proportionate risk are by far the most significant and innovative (Prieur 1996, 14).

The *Charter* initially addresses the issue of informed consent in its section dedicated to informed consent given by the patient. First and foremost, the "health care worker should have the express or tacit consent of the patient" before proceeding with any intervention (n. 72).[47] To buttress this assertion, the *Charter* recalls the words of Pius XII, according to which the health care worker "does not have a separate and independent right in relation to the patient. In fact, he can act only if the patient explicitly or implicitly (directly or indirectly) authorizes him" (Pius XII 1957, 1031). These words have been confirmed by Pope John Paul II, who emphasized that "the doctor has only that power and those rights which the patient himself gives him" (John Paul II 1980a, 1009). Thus, the Pope and the *Charter* leave no doubt as to whether or not the subject's self-determination ought to be respected. Nevertheless, the *Charter* never employs terms like "self-determination" and "autonomy," thereby distinguishing itself from the type of argumentation deeply rooted in present-day mainstream secular bioethics.

Second, the *Charter* depicts the relationship between patient and physician as "a human one: dialogic, non-objective." Viewed in such way, the relationship in question provides the basis for informed consent. The words of Pope John Paul II invoked in the document explain that the patient is "a responsible person who should be called upon to share in the improvement of his health and in becoming cured. He should be given the opportunity of personally choosing, and not be made to submit to the decisions and choices of others" (John Paul II 1982b, 673). A few paragraphs later, the *Charter* applies this requirement to the realm of experimentation and goes on to say that man "because of his unique dignity can be the subject of research and clinical experimentation with the safeguards due to a being with the value of a subject and not the object" (n. 75). This personal approach, resistant to objectivization, makes it clear that the subject or patient is not the one *on whom* but rather the one *with whom* medical procedures are to be carried out. Thus, the perspective utilized here turns out to harmonize with the neglected legacy of the Nuremberg Code. This legacy, as explained by Herranz and recalled earlier in this study, has never been fully used in human-rights documents. By way of reminder, one feature abandoned by the post-Nuremberg documents gives prominence to the fact that informed consent translates itself into ethical responsibility rather than merely legal requirements, avoidance of punishment, or civic duty, and embraces both the subject and the researcher. By stressing the relevance of the "dialogic, non-objective" relationship for informed consent, the Magisterium not only places itself in accord with this provision but also enriches and deepens it by recourse to both philosophical and theological arguments. Likewise, a second later-abandoned feature of *Nuremberg*, highlighting not only the researcher's but also the subject's responsibility for the research project, is addressed here. The *Charter* implicitly denies the possibility of the withdrawal of subjects from the research project merely on the basis of felt desires or whims. Mature responsibility certainly calls for more and has chiefly to do with the moral character of both researcher and subject.

Third, when bringing up the topic of presumed consent, the *Charter* distinguishes between "the patient who is in a condition to know and will and one who is not" (n. 73). In the former case, presumed consent is never licit, whereas it might be legitimate in the latter situation. To help clarify this matter, the document employs two principles, namely that of therapeutic trust and that of responsibility for health care. The principle of therapeutic trust finds its application in a situation where "there is a temporary loss of knowing and willing" on the part of the patient and points to the "original confidence with which the patient entrusted himself to the health-care worker." If the loss of "knowing and willing" proves permanent, "the health care worker can act in virtue of the principle of responsibility for health care, which obliges the health care worker to assume responsibility for the patient's health"

(n. 73). The health care worker's responsibility does not preclude the responsibility of others, in particular of the patient's relatives who "should be informed about ordinary interventions, and involved in the decision making when there is question of extraordinary and optional interventions" (n. 74). Thus, the ethical perspective ingrained in the subject-researcher relationship and evoked in the previous paragraph has been strengthened with the essential components of trust, confidence, and the extension of responsibility not only to the research team and subject but also to the patient's relatives.

Although all of the above-listed provisions have been placed in the context of medical care, they also apply to experiments involving human subjects. Still, the *Charter* offers an additional section covering the domain of experimentation. Immediately at the outset of that section, the distinction between research and experimentation is introduced, according to which the latter uses and tests on a subject the findings of the former. Both research and experimentation need to conform fully to "the good of the person." In addition, research must contribute to the promotion of human well-being (n. 76).

In order for the researcher to safeguard the good of the subject, both informed consent and risk factors need to be taken into account. The subject ought to be provided with information concerning the purpose of the experiment and its potential risks so that he or she can make a free and informed decision (n. 77). Regarding the risk factors, the *Charter*, mindful of the sober words of Pope Pius XII signaling that the exclusion of all risk "would paralyze all serious scientific research and would quite often be detrimental to the patient" (Pius XII 1952, 788), nonetheless states that "there is a threshold beyond which the risk becomes humanly unacceptable" (n. 78). The gauge by which this threshold is to be measured can be found by referring to "the inviolable good of the person" that forbids the subject "to endanger his life, his equilibrium, his health, or to aggravate his illness" (John Paul II 1986, 1183). In order to decrease risk and support the subject's safety, "the *testing* of new pharmaceutical products or of new techniques should first be done *on animals* before they are tried on humans" (n. 79, emphasis in text).

As laid out in the first chapter, the human-rights documents dealing with experimentation on human subjects stipulate in principle that the risks involved must be offset by hoped-for benefits. Furthermore, in order for the research project to be justified and approved, these elements must remain in a favorable ratio. Similarly, the *Charter* employs the principle of proportionate risk, "that is, of due proportion between the advantages and foreseeable risks" (n. 79). The principle works differently depending on whether exercised in therapeutic or nontherapeutic research. In the former case, even higher-risk procedures might be employed if all safe remedies have been exhausted and "great respect" for the subject is preserved (n. 80). In the latter instance, only a narrow scope of risk may be authorized, which is to say that it cannot exceed "the risk of any insubstantial physical impairment" (John Paul II 1985, 1009). In effect, even if informed consent were to be secured for an experiment involving the risk of substantial physical impairment and thus a considerable threat to subject's integrity, such consent would not validate the potential experiment.

3.1.2 The Ethical and Religious Directives for Catholic Health Care Services

Published by the United States Conference of Catholic Bishops, the *Ethical and Religious Directives for Catholic Health Care Services* is said to be "the main compendium of Catholic teaching for Catholic health care institutions and professionals in the United States" (Ashley and O'Rourke 2002, ix). The *Directives* have been revised and updated several times. Here, I examine their latest version, the fourth, issued June 15, 2001.[48]

In accord with the *Charter for Health Care Workers*, the *Directives* instruct that free and informed consent is required for medical treatment and procedures (n. 26) as well as for "medical or genetic experimentation even if it is therapeutic" (n. 31) Yet, while the *Charter* locates the theme of the professional-patient relationship in the section devoted to informed consent, the *Directives* do the reverse. By placing the matter of consent within the unit entitled "the professional-patient relationship" and by stating that free exchange of information between professional and patient results from their relationship (part III, introduction), the US Bishops' document puts an even stronger emphasis on the essential nexus existing between informed consent and the medical relationship which "requires, among other things, mutual respect, trust, honesty and appropriate confidentiality."[49]

Indeed, only when these provisions are met can genuinely free and informed consent be attained. The provisions by their very nature prohibit any element of threat, dread, coercion, or undue influence and thus uphold the patient's freedom. Moreover, they require that all necessary information be given to the patient. By affirming that "neither the health care professional nor the patient acts independently of the other" (part III, introduction) the *Directives* underscore that both participate and cooperate in the same project. Consequently, they both bear moral responsibility for that project.

After depicting the medical relationship that renders possible the free exchange of information as a prerequisite for securing valid informed consent, the *Directives* go on to specify that both medical and moral information should be available for the patient (n. 28). What is more, "each person or the person's surrogate should have access to ... counseling so as to be able to form his or her conscience" (n. 28).[50] This statement deserves greater attention.

To begin with, no human-rights document examined in the first chapter of this study explicitly refers to counseling as an essential part of the process of obtaining informed consent. At one point, the CIOMS Guidelines use the word "counseling" when commenting on Guideline 6, which refers to obligations of sponsors and investigators during the process of securing consent. The sentence in question reads: "potential subject should receive counseling about risks of acquiring a disease unless they take precautions." The counseling thus referred to is used in the context of a purely technical risk-related requirement which allows for treating it merely as the equivalent to providing information. No further specification is offered. By contrast, the *Directives* see counseling as related to the formation of conscience, which denotes considerably more than simply delivery of technical information.

It is noteworthy that none of the human-rights documents dealing with informed consent examined in the first chapter recognized the need for formation of conscience. In fact, none of these documents ever used the word "conscience" except for the Declaration of Helsinki whose second paragraph reads: "it is the duty of the physician to promote and safeguard the health of the people. The physician's knowledge and conscience are dedicated to the fulfillment of this duty." Still, the Declaration never refers to the conscience of the patient, which is a substantial shortcoming. At a more elementary level, what is lacking here in the light of the *Directives* is an explicit reference to the objective good of the patient, which is not reducible to his or her subjective preferences. These preferences do not define by themselves the good of the patient; rather it is the real, genuine good that shapes the preferences. This is precisely the role that conscience ought to play, a fact of which the *Directives* are clearly aware.

As I have shown previously, the human rights documents in general—the Nuremberg Code and to a lesser degree the Belmont Report are exceptions—consistent with the methodology they adopted, endorse the professional's following even what is merely a subject's fleeting caprice or fad out of concern for safeguarding the research subjects' autonomy. Any other approach would most likely be considered a type of paternalism, posing the threat of undue influence or coercion and thus not adequately respectful of the patient's or the subject's autonomy.

Putting aside the question of its diverse definitions and types, paternalism, even if identified in purely negative terms or evoking exclusively pejorative associations, may be justified in certain situations.[51] Even when not fully justified, it still needs to be distinguished from attitudes or behaviors that admittedly resemble it but in fact vary considerably from it. The common denominator of all forms of paternalism can be reduced to a certain kind of imposition bringing about the limitation of freedom or the overriding of one's preferences (cf. Beauchamp 2004, 1984). This characteristic, however, does not pertain to something that remains within the bounds of a legitimate expression of trust for the good of the patient or research subject even if this good happens to be in opposition to his or her preferences.

Now, what is included in the expression of trust for the patient's or subject's good ought to be described as proposing rather than imposing certain solutions. To put it another way, instead of imposing some sort of good, a responsible professional or perhaps someone else legitimately involved in the decision-making process, proposes to the patient or subject the consideration of his or her objective good, which, even if not yet fully known, may be discovered, rediscovered, or simply better understood by both professional and patient/subject as an effect of their dialogic interaction based on mutual trust. It is true that on the one hand the counseling thus conceived may yield nothing more than a pure exchange of information, or it could equally well move to the opposite extreme and become a form of unjustified paternalism. How to avoid the danger of these two extremes? Unlike the human-rights documents, the *Directives* offer a simple and firm answer: by recourse to conscience. Thus the link between formation of conscience and the frontiers of legitimate counseling aimed at the true good of a patient becomes critical in the decision-making process in both therapy and research settings.

A constant effort to reach a better knowledge of one's real and objective moral good touches on the most profound dimensions of one's moral character and its enhancement. Persons become good not in the manner of monads; they need formation, presupposing a meaningful interaction with both God and other persons. This is the idea that the *Directives* bring up when they refer to conscience. By introducing a profound notion of medical relationship they reach far beyond the somewhat simplistic view, fearful of paternalism, that is characteristic of the human-rights documents, and point instead to the fact that counseling related to formation of conscience proves to be an indispensable element of every truly genuine informed-consent process. Needless to say, counseling in the experimental context, although perhaps not as critical as its health-care counterpart, also plays a crucial role at virtually every stage of research, especially when new issues, not fully realized or foreseen before the experiment began, arise during its course. Given the reality of the experiments on humans carried out by the Nazis during World War II as well as the abusive research done in the United States in the recent past, the call for a well-formed conscience of those involved in research projects must be attended to with the highest urgency and relevance.

In addition to conscience-related issues, the *Directives* provide for exceptions to informed consent which are allowed "in an emergency situation when consent cannot be obtained and there is no indication that the patient would refuse consent to the treatment" (n. 25). In other words, consent can be presumed only in an emergency situation, a factor not explicitly expressed in the *Charter for Health Care Workers*. The *Directives*, in turn, do not mention the principles of therapeutic trust and the responsibility for health care introduced by the *Charter*. Yet, they can be said to tacitly assume these principles and complement their application by referring to an emergency situation as the condition sine qua non for the defensible presumption of consent. Furthermore, while reference to the patient's conscience plays a principal role with respect to the medical and moral information that should be made available to the patient before securing informed consent, it is the conscience of the professional that becomes critical as regards emergency situations in which presumed consent may be necessary.

3.1.3 Conclusion

In addition to the documents just presented, a number of Magisterial texts exist, which, although never making reference to informed consent explicitly, nevertheless fundamentally condition its understanding. An example of such a passage would be an excerpt from the Pastoral Constitution on the Church in the Modern World, *Gaudium et Spes,* of the Second Vatican Council. When addressing the topic of human freedom the Constitution reads: "Man's dignity . . . requires him to act out of conscious and free choice, as moved and drawn in a personal way from within, and not by blind impulses in himself or by mere external constraint" (n. 17).[52] Interestingly, the Constitution does not pronounce merely on "free choice" but rather on "free and conscious" choice, thus relating the matter more strictly to human responsibility and conscience.

Within this perspective, human freedom proves inextricable from the human dignity that provides the basis for it and ultimately explains its meaning. Therefore, the principle of informed consent, with its apex being the free choice of patient or research subject, cannot be seen as divorced from the truth on human dignity and freedom (see Griese 1987, 154–57). This conclusion finds firm corroboration in the *Catechism of the Catholic Church*, which declares: "research or experimentation on the human being cannot legitimate acts that are in themselves contrary to the dignity of persons and to the moral law. The subject's potential consent does not justify such acts." At a still more rudimentary level, "experimentation on human beings does not conform to the dignity of the person if it takes place without the informed consent of the subject or those who legitimately speak for him" (n. 2295).[53]

In relation to this, it seems appropriate now to identify one additional significant difference between the human-rights documents on the one side, and the Magisterial teaching on the other. Put simply, the latter, in opposition to the former, provides a moral benchmark by which to interpret and consistently apply its principles. Regrettably, the human-rights documents lack such a benchmark. Still, it must be admitted that the Nuremberg Code differs from other human-rights documents in this regard. By building on the basic moral content on which all agree, the Code earnestly attempts to ground its principles in the theory of natural law. Consequently, it directly addresses the ethical dimension of informed consent as well as the personal moral responsibility of both researcher and subject, the responsibility which is unalienable by its very nature. Having said this, it can still be argued that the dearth of more explicit and categorical references to objective human good and human dignity makes it hard to find in the Code a cohesive yardstick for interpretation. Moreover, as already mentioned, because of the historic context of its origination, its purpose, and its concise content, the span of research which *Nuremberg* actually covers remains debatable.

Conceivably, the CIOMS Guidelines with their four principles, echoing those brought into bioethics by Beauchamp and Childress and articulated in the Guideline's introductory section, could also be considered by some as self-explanatory. Although seemingly tenable, this supposition does not hold for two reasons. First, the Guidelines have almost nothing to say about the principles' meaning, scope, and structure. Second, the principles themselves relate very loosely, if at all, to the Guidelines' subsequent content. In fact, no direct link between them and the ensuing requirements has been demonstrated. Nor does the document show at any point how to implement the principles in the practical research context. Indeed, while perusing the Guidelines, not only does one not find an explanation of the principles' meaning but one also confirms the suspicion that there is no connection whatsoever between them and the contents of the Guidelines which they precede. In effect, one can hardly regard the four principles as a useful key to ethical interpretation. Perhaps recourse to the professional bioethics literature might be of some help, but given the fact that scholars differ in their views on the four principles, the question about the ethical meaning intended by the authors of the Guidelines, which is to say the meaning that should be applied to the Guidelines' content, remains unresolved.

Therefore, in my opinion, all of the documents examined in the first chapter, with the partial, albeit notable, exclusion of the Nuremberg Code, although replete with technical and procedural comments, offer no reliable ethical yardsticks by which to interpret and apply their provisos. By saying this, I do not mean that the documents in question are totally free of any sort of ethical or moral terminology. Such a conclusion would certainly not do them justice. What I mean is that despite incorporating some ethical concepts and ideas, the human-rights documents only minimally explain the meaning and applicability of these concepts and ideas. As a matter of fact, they typically point to these concepts in a liberal, loose, and relatively non-committal manner. The necessary generality of the various provisions found in these documents renders them nebulous and potentially puzzling unless clear criteria for interpretation are given. I contend that the substantial deficiency of human-rights documents lies in the absence of such criteria. In consequence, they can be interpreted in a variety of ways, with their provisions understood and applied differently by people of diverse intellectual bents and moral preferences. Not surprisingly, the conclusions of such interpretative efforts may stand in stark contrast to the original intentions and objectives of those who authored the human-rights documents. I have already briefly illustrated this dilemma when examining the Belmont Report, with its wording subject to different interpretations of the notion of autonomy.

While the absence of standards of moral interpretation in the human-rights documents exposes them to entirely subjective or even arbitrary interpretations and applications, the Magisterial teaching on informed consent, although not as technically detailed, offers solid anthropological and ethical grounds on which to elaborate on concrete moral solutions regarding, for instance, the realm of experimentation on human subjects. Concerning its moral implications, the Magisterial teaching firmly recognizes the foundational character of rational human nature. By referring to natural law and human dignity as both the foundation of moral principles and the ultimate criteria by which to seek and measure the objective good of patient or research subject, the Magisterium affirms the personal status of every human being and safeguards the subject's integrity and well-being at every stage of life—a truly novel approach in comparison with that of the human-rights documents.

Moreover, the account of human freedom as based on natural law helps to show clearly the intrinsic relationship of this freedom to the responsibilities and obligations binding all engaged in the research enterprise regardless of their frame of mind and preferences. By appreciating the role of conscience, entirely passed over by the human-rights documents, the Magisterial documents not only establish boundaries for a legitimate counseling free of coercion and undue influence but also draw attention to the personal character of every decision involved in the informed-consent process. "Personal," as perceived here, does not denote "private," "anonymous," "individualistic," or "separated" (the meanings sometimes suggested or tacitly conveyed by the human-rights documents) but rather ethical or moral in the sense of meeting objective criteria determined by rational human nature and the human dignity which is intrinsic to it. Both recognition and application of those criteria presuppose formation of conscience which in turn touches upon the moral character of subjects and investigators—again, subject matter completely omitted by the human-rights texts.

To put matters in simpler terms, in the view of the Magisterium, what persons should do ultimately does not revolve around the set of guidelines to which they more or less intentionally subject themselves but rather rests upon who those persons are. In short, the quality of informed consent and in general the safety of research subjects hinges not so much on the quality of the guidelines or other documents as on the quality of those who implement them in research. With respect to this moral quality the Nuremberg Code appears again to coincide with the Church's teaching. It was for a reason that the Code conceded in its first principle that "the duty and responsibility for ascertaining the quality of the consent rests upon each individual who initiates, directs or engages in the experiment. It is a personal duty and responsibility which may not be delegated to another with impunity." In other words, the quality of consent hangs on the personal duty whose execution in turn relies on the moral quality of the one who bears this duty. It follows that the moral quality of all involved in the experiment remains in direct proportion to the quality of consent. Obviously, this proportion works differently in case of a subject about to decide whether or not to give informed consent and of a physician or perhaps some other persons, such as a healthcare team or family members, who assist the subject. In regard to that, the Magisterial documents reach beyond the Nuremberg perspective and point to the relevance of a dialogic trust-based relationship between the investigator and the research subject. This relationship has been conceived as a guarantor of obtaining valid, free and informed consent. Once again, the human-rights documents, although referring to this relationship, still put much less emphasis on its ethical weight and on its indispensability for informed consent.

As can be seen throughout the present section, the ethical incisiveness of Magisterial documents, paired with their theological and philosophical lucidity, places them closer to the Nuremberg Code than to the remaining documents examined in the first chapter. The apparent ethical concurrence between the Nuremberg Code and the Magisterial teaching becomes even more impressive when observed through the lens of Hippocratic tradition. As noted earlier, in the instance of the *Charter for Health Care Workers*, that concurrence can be regarded as "a synthesis of Hippocratic ethics and Christian morality" (Honings 1996, 48), "for Hippocratic ethics, as for Christian morality . . . the life of each human being is a value which cannot be called into question—it must be defended and watched over. In a word, it must be served" (Honings 1996, 48). When investigator and researcher become servants of life, their mutual relationship undergoes a transformation so that it can be characterized as "an encounter between trust and conscience" (Honings 1996, 49). Without this perspective, one remains incapable of understanding either Christian morals or the Hippocratic ethos that, when detached from reciprocal trust and conscience joined in unreserved deference to human life, may indeed deteriorate into the worst forms of paternalism.

3.2 Relevant Philosophical and Theological Approaches

This section discusses major philosophical and theological accounts of informed consent. The literature on this matter is vast and therefore, due to space constraints, in what follows I will confine myself to the examination of the most relevant positions. I will begin by scrutinizing Paul Ramsey's views, which are embedded in the biblical norm of fidelity to covenant. Next, I will turn to Edmund Pellegrino's and David Thomasma's defense of virtue. My analysis thereof will emphasize the virtues of prudence and integrity, followed by a description of the way in which they sway the informed-consent process. In the third part, I will address an approach epitomizing principlism, an influential current in present-day bioethics. Among its most prominent advocates belong such authors as Tom Beauchamp and James Childress. Their views, along with those of Ruth Faden, seek to ground informed consent on the prima facie principle of respect for autonomy. Finally, I will close the current section by discussing Germain Grisez's approach to this issue, emphasizing the role of conscience and prudence in his approach. In addition, I will present his clarification of the significant-risk criterion, which is critical for informed consent, including its proxy type.

3.2.1 Paul Ramsey

In his book, *The Patient as Person,* Paul Ramsey offers an analysis of informed consent perceived as being "at the heart of medical care" and defined in terms of "a joint adventure between patient and doctor" (Ramsey 2002, 11). Ramsey has also depicted the principle of informed consent as "a canon of loyalty expressive of the faithfulness-claims of persons in medical care and investigation" (Ramsey 2002, 10). The faithfulness-claims, in turn, are placed upon the researcher by the patient or subject merely by reason of their humanity, and as such constitute "the ground of the consent-rule in medical practice" (Ramsey 2002, 8). The reference to humanity, although it seems to resemble the Magisterial teaching on the normative, objective meaning of human nature, nonetheless conveys a different message.

Ramsey holds along with Karl Barth that "covenant-fidelity is the inner meaning and purpose of our creation as human beings" (Ramsey 2002, xlv) and in this sense covenant-fidelity is natural to us. Thus, in place of the ontologically oriented account of human nature characteristic of the Catholic approach which, although congruent and harmonious with Scripture yet reaches beyond it, Ramsey introduces one based exclusively, and somewhat narrowly, on the biblical norm of fidelity to covenant. This effort corresponds to a critique of natural law to be found in Ramsey's classic work *Basic Christian Ethics*, according to which natural moral law is not the Christian source par excellence of ethical insight and hence it cannot be utilized as the tenet on which to build Christian ethics (Hauerwas and Long 1993, xix). Nevertheless, despite taking a strong stand against natural law, Ramsey has never specified what his understanding of its content was (Hauerwas and Long 1993, xx).[54] Instead, he chooses to bind Christian love with sacredness of life as the basis for ethical reflection (see May 2001, xxxix–xliii). Christian love, that in his view is supposed to transform natural law (Hauerwas and Long 1993, xix),[55] finds its roots in "God's covenanting activity with humanity. Love and covenant should not be separated, for they are but two ways of indicating God's care for God's creation" (Hauerwas and Long 1993, xv; see also Jonsen 2003, 48). It follows that informed consent, being the expression of covenantal fidelity, cannot depart from love. Indeed, the canon of loyalty presupposes love and regards it as an inextricable element.

As far as the second source of Christian ethics, sacredness of life, is concerned, Ramsey's understanding also does not entirely tally with Catholic thought. Ramsey stresses that human life is sacred not in itself but inasmuch as it draws from the sanctity of God. In *Basic Christian Ethics*, he refers to "all human worth as derivative, not inherent" and notes that "Christian interpretations of man's dignity affirm something about man in relation to God, not just something about man per se" (Ramsey 1993, 277).

The possible paths of critique from the Catholic standpoint of such a formulation could be directed toward the fact that human life itself has never been capable of existing as separated from God, its creator, constant sustainer, and ultimate end. From the very first moments of its existence, life proves derivative of God, which means nothing else than that it is sacred both in itself and as derivative of God. To put it another way, life in itself is derivative of God, which is to say the inherent worth of human life does not stand in conflict with its derivativeness of God. Hence, the seeming rift between these two aspects of life proves rather artificial, or at best merely conceptual, not real. Ramsey, however, sees the notion of human dignity as not only vulnerable to reduction itself but above all as a downgraded form of sanctity of life (Ramsey 2002, xlvi), a claim neither satisfactorily elucidated by Ramsey nor consonant with the Catholic perspective. These difficulties notwithstanding, as some commentators have pointed out, in *Patient as Person* Ramsey discarded the insights on the sacredness of human life outlined above and embraced a position more defensive of the inviolability of life, especially that of patients and research subjects (see e.g. Hauerwas and Long 1993, xxiii).

Although it is beyond my task to verify the accuracy of this speculation, it is true that *Patient as Person*, in spite of its recurrent nebulousness or even elusiveness, nonetheless ought to be appreciated as an insightful and stimulating contribution to medical ethics with a prominent place given to the worth of humans as patients or subjects. Ramsey captures medical ethics as "consonant with the ethics of a wider human community." Thus, "canons of loyalty to patients or to joint adventurers in medical research are simply particular manifestations of canons of loyalty of person to person generally" (Ramsey 2002, xliv–xlv). It follows that by the mere fact of being a person one bears the responsibility of loyalty towards other persons. Indeed, "we are born within covenants of life with life" (Ramsey 2002, xlv), as Ramsey verbalizes it. Covenant implies fidelity, which in turn strictly relates to trust.[56] Certainly, both have a decisive share in the process of informed consent. Ramsey sees fidelity as "the bond between consenting man and consenting man." Remarkably, it is not only the patient or subject who consents; it is also the investigator. In other words, two consents of two persons need to meet in the process of securing consent, thus rendering fidelity upon this reciprocal event. By the same token, Ramsey underscores the role of trust, whose meaning can be grasped by the maxim that reads: “act always so as not to abuse trust” or “act always so as to exhibit faithfulness, to deserve and inspire trust” (Ramsey 2002, 8, note 6).

Furthermore, it can be argued that, although plainly Christian, Ramsey’s analysis, employing as it does the biblical notion of covenant, is closely aligned with the Jewish tradition. By emphasizing the role of love as underlying the entire reality of Christian life it also holds a well-defined place within Christian ethics of Protestant provenance. In addition, despite the differences from the Catholic approach noted above, Ramsey’s account can be reconciled, to a degree, with Catholic moral doctrine. Probably the best illustration of that can be found when Ramsey turns to the nature of medical ethics. After expressing his reservations regarding a purely consequentialist ethics and offering notes of disapproval on contractual and situational ethics, Ramsey goes on to acknowledge the value of duty and suggests that it is the moral character of a researcher and research subject rather than any other component on which medical ethics in general and informed consent in particular rest (see Jonsen 2003, 50–51). Moreover, Ramsey does not hesitate to admit that “the practical application of the requirement of an informed consent is always the work of prudence, which does not mean caution but practical wisdom in the appraisal of cases and specific situations” (Ramsey 2002, 3).

The linking of prudence with practical wisdom in Ramsey’s work traces back to St. Augustine, whose writings are copiously cited in *Basic Christian Ethics* (see Ramsey 1993, 202–13). St. Thomas Aquinas’ insights are also critically referenced on more than one occasion. Elaborating on the thought of Augustine and Aquinas brings about the conclusion that “virtues may become forms of the love of God” (Ramsey 1993, 207–208), which sheds additional light on the relation existing between love and the covenantal fidelity that now turns out to encompass virtues, too. Consequently, the practical application of informed consent cannot depart from Christian love not only on account of its inextricable ties to covenant but also because it presupposes and requires virtue. To put it another way, for Ramsey to admit that the application of informed consent is the work of prudence means that it is primarily and principally the work of love.

In spite of not finding himself in full accord with both Augustine and Aquinas, Ramsey seems to connect, at least to some extent, with the Catholic understanding of prudence as the first of the cardinal virtues, the one that, according to the *Catechism of the Catholic Church*, “disposes practical reason to discern our true good in every circumstance and to choose the right means of achieving it.” Thus perceived, prudence not only “guides the other virtues by setting rule and measure” but also “immediately guides the judgment of conscience.” Therefore, “with the help of this virtue we apply moral principles to particular cases without error” (n. 1806). Undoubtedly, the last sentence is in accord with Ramsey’s position: it is by means of the virtue of prudence that the moral principle of informed consent can be successfully applied in concrete situations. But what is even more significant from the perspective of informed consent is the relationship between the virtue of prudence and conscience contained in the quote from the *Catechism* and clearly recognized by broad Catholic moral theology.[57]

To sound out to what extent, if any, Ramsey was aware of this relationship, constitutes a formidable task and indeed, one beyond the goal of this study. In *Basic Christian Ethics,* conscience has been referred to sporadically with neither its meaning nor its structure explained. Other works of Ramsey, with *War and the Christian Conscience* (Ramsey 1961) at the forefront, are not of much help either. Nonetheless, regardless of his convictions on conscience and despite admitting that "the medical profession should no longer believe that the personal integrity of physicians alone is enough" (Ramsey 2002, li), Ramsey holds this integrity in high esteem and contends that "the investigator should strive, as Aristotle suggested, to hit the mean of moral virtue or excellence by 'leaning against' the excess or the defect to which he knows himself, individually or professionally, and mankind generally in a scientific age, to be especially inclined" (Ramsey 2002, 10). One of such excesses or defects occurs when a researcher happens to be strongly attached to the research he or she carries out so that it is the research itself rather than its subjects that becomes of an utmost importance. Seen from this angle, the requirement for consent, "is no substitute for—and indeed there can be no substitute for—the wisdom and moral integrity of the medical practitioner" (Ramsey 2002, 8).

Although he does not convert the consent requirement into an absolute and he speaks of *reasonably* free and informed consent (emphasis added), and despite being acutely aware that entirely free and informed consent presents an obligation that cannot always, if ever, be met,[58] Ramsey situates informed consent at the heart of medical practice. Defined in intrinsic relation to covenantal love, based on fidelity and trust, and classified as a canon of loyalty, informed consent proves workable when virtue remains recognized, valued, and encompassed in medical practice.

3.2.2 Edmund Pellegrino and David C. Thomasma

Recognition of virtue in medical ethics never came to the fore as vigorously and keenly as in the work of Edmund Pellegrino and David Thomasma. In their book *The Virtues in Medical Practice* the following eight virtues have been singled out as particularly pertinent to the subject: fidelity to trust, compassion, phronesis (prudence), justice, fortitude, temperance, integrity, self-effacement (see Pellegrino and Thomasma 1993, 65–161). The accounts of the virtue of prudence and the virtue of integrity best serve this chapter's purpose because of their intimate relationship to the field of research on humans in general and informed consent in particular.

Called "medicine's indispensable virtue," prudence (*phronesis*) has been firmly present in ethical reflection since Aristotle, who understood it as practical wisdom, which is to say "the capacity ... in a given set of circumstances, to discern what moral choice or course of action is most conducive to the good of the agent or the activity in which the agent is engaged." From this it follows that prudence not only enables a moral agent to "discern which means are most appropriate to the good in particular circumstances" but also allows for attaining the good (the end) "for the agent and the work in which he or she is engaged" (Pellegrino and Thomasma 1993, 84; see also Pellegrino 2001c, 126). Owing to these two features, essential for any ethical endeavor, prudence since antiquity has been considered the cornerstone virtue, the critical nexus among the intellectual, moral and—as was the case of Thomas Aquinas along with some other medieval and post-medieval thinkers—supernatural virtues. The link to the supernatural helps one realize that the notion of prudence evolved throughout centuries: in a nutshell, Aristotle's *phronesis* gained the additional yet no less relevant characteristic consisting in "the appreciation of a faith commitment" (Pellegrino and Thomasma 1993, 85).

Before they demonstrate how prudence works when applied to concrete medical situations, Pellegrino and Thomasma provide for two ends of medicine constituting its good, that is, "that to which the virtues of the physician qua physician should be directed." Overall, "the ultimate end is the health of individuals and society, while the more proximate end is a *right* and *good healing* action for a specific patient" (Pellegrino and Thomasma 1993, 86, emphasis added).[59] The *right action* refers to what is suitable and scientifically sound from the medical point of view and thus bears upon purely medical good. By describing the action as *good*, one takes into consideration the subject and his or her own "values, lifestyle, aspirations, religious beliefs and so on." Consequently, for a medical action to become morally good, the one who performs that action needs to reach beyond what is merely right from the scientific and medical perspective and seek the good as perceived by the patient.[60] Moreover, beyond being right and good, the action in question is to be healing. Our authors distinguish between curing and healing, assigning the former to disease and the latter to the patient. By doing so, they capture healing "in its broadest sense." Certainly, "even if it is not possible to cure or contain the disease, healing can occur if the patient is assisted to cope with her illness" (Pellegrino and Thomasma 1993, 86).

No matter whether attended to in a therapeutic or nontherapeutic milieu, the two ends of medicine require of a physician/researcher specific habitual dispositions or traits of character, that is, virtues, among which "prudence plays a special role because of the nature of medicine as an activity . . . requiring a fusion of technical competence and moral judgment . . . Prudence is therefore both an intellectual and a moral virtue in medicine, as it is in moral encounters generally" (Pellegrino and Thomasma 1993, 86–87).[61]

At this point it is time to ask how prudence affects the informed-consent process. One example offered by Pellegrino and Thomasma in the context of the virtue of honesty seems particularly useful here. As this study has demonstrated up to this point, the adequate and unbiased disclosure of information makes up one of the most germane facets of informed consent. In practice, both the information itself and the way it is volunteered are crucial. As Pellegrino and Thomasma put it, "experienced clinicians . . . assert, with considerable accuracy, that they can get almost any decision they want, depending on how they use their authority and knowledge to present the choices" (Pellegrino and Thomasma 1993, 87).

Does not this description perfectly fit the research settings? It does, because in order to have unhindered access to the group of subjects on whom they want to experiment or to convince the potential subjects to join in the research project they run, the investigators may resort to employing unfair and dishonest tactics. Such tactics may include, for example, conscious, albeit typically covert, manipulation; a selective approach to scientific facts; or simple deception. "When does obtaining consent become coercive? When should the physician try to persuade, and when is he unduly influencing the decision? Is it not tantamount to moral abandonment not to advise the patient on the basis of our best estimate of the interest of the patient? Yet, if we express our own preferences, do we not subtly overmaster the vulnerable patient?" (Pellegrino and Thomasma 1993, 88). Pellegrino and Thomasma argue that satisfactory answers to these and similar questions can be found only by recourse to the virtue of prudence. A prudent physician/researcher is one whose moral character not only recognizes what is right and good but who is also habitually inclined to act in accord with what right and good demands so as to achieve good by divining and harnessing the most appropriate means leading to it (see Pellegrino and Thomasma 1988, 115).

Besides prudence, another high-priority virtue pivotal to the informed-consent process and examined by Pellegrino and Thomasma is that of integrity. As far as its etymology is concerned, integrity denotes completeness, wholeness, and unity (Pellegrino and Thomasma 1993, 129). Thus construed, it "defines . . . the nature of the individual who integrates all of the virtues" (Pellegrino and Thomasma 1993, 127). Still, the concept of integrity can be examined at two levels: "the integrity of person" that is of the patient, subject, or physician, and "being a person of integrity." While the first level "is expressed in a balanced relationship between the bodily, psychosocial, and intellectual elements" (Pellegrino and Thomasma 1993, 131) of someone's life, the second pertains to the moral wholeness and soundness of a physician/researcher to whom another is entrusted.

In medical practice these two aspects interact and have an impact on each other. Our authors put the matter as follows: "the physician must be a person of integrity if the integrity of the person is to be safeguarded and healed when it is dis-integrated by illness" (Pellegrino and Thomasma 1993, 132). Naturally, this has a significant effect in the area of experimentation on human subjects, in which integrity of both researchers and subjects is frequently endangered. In some cases, when subjects participating in research are healthy, dis-integration may occur either during the experiments or as their result. Thereupon, taking pains, no matter how great, to assure the safety and the integrity of the research enterprise constitutes a matter of the highest importance. Ultimately, "it is the character of the investigator that determines the moral quality of research" (Pellegrino and Thomasma 1993, 133). By the same token, "the moral integrity of the investigator who obtains the consent is what assures the moral validity of the consent procedure" (Pellegrino 2001a, 194). Aside from the obvious fact that it is the investigators who weigh, appraise, and navigate through various intricacies and subtleties inherent to carrying out the experiments and who apply the principles and rules, along with deciding which one takes precedence in case of conflict, no research undertaking escapes a certain amount of moral gray area within which each subject, no matter how well educated and determined in exercising his or her autonomy, relies significantly not only on the knowledge of the researcher but first and foremost on his or her integrity and on integrity's subsidiary: intellectual honesty (Pellegrino and Thomasma 1993, 127). In other words, neither principles nor guidelines, nor even informed consent itself comprises the ultimate protection of subjects. In fact, all comes down to the moral character of the investigator within which virtues occupy the central position (see Pellegrino 2001a, 193–94). Consequently, securing valid and meaningful consent requires a set of virtues among which integrity and prudence take precedence.

The prominence of moral character has become even more evident recently as scientific research has expanded beyond scholarly activity and reached out to industry and trade. In line with Pellegrino and Thomasma's explanation, "the values of scholarship—seeking truth, reporting honestly, sharing information, restraining bias, and protecting the freedom to choose research subjects—are brought into conflict with the values of industry, commerce and business . . . The industrial model makes research a commodity like any other, an item produced for its exchange value, for what it yields in promotion, tenure, lighter teaching loads, a larger laboratory, more support dollars . . ." Under such circumstances, "the senior investigator becomes an entrepreneur and a corporate executive whose success is measured in the number of works published, grants received, committee memberships, prizes and the like" (Pellegrino and Thomasma 1993, 136). In consequence, the transmutation of the nature of scientific research not only fuels self-interest[62] which undermines investigator's integrity and results in different sorts of misconduct that become extremely perilous when human subjects are involved, but also compromises truth, the principal value of science.[63] Building on Alasdair McIntyre's analysis delineating virtues as "acquired traits that are necessary to attaining the goods *internal* to a practice," Pellegrino and Thomasma go on to say that since "the good internal to practice of research is truth . . . the virtues of the scientists are those that enable the scientists to attain truth which is the aim of research" (Pellegrino and Thomasma 1993, 135). Although our authors do not mention this explicitly, it appears safe to contend that, by analogy to the distinction between right and good action drawn above in the context of two ends of medicine, truth, being the aim of research, consists not only in scientific or technical knowledge but also in comprehension of the human condition of research subjects. This is why in the selection of subjects, for instance, one should care not only about the scientific criteria qualifying them for participation in a particular project but also about their humanity which, besides economical and social factors, embraces philosophical and anthropological truth about human nature: its structure, inclinations, potentiality, and limits.

The virtues making it possible for an investigator to reach and sustain truth are, for example, objectivity, critical thinking, and freedom from bias. By integrating them and applying them with prudence to concrete experimental situations, the researcher remains in service of truth. In contrast, when goods external to the practice of research, such as exorbitant ambitions, personal profit, or prestige, take over truth and stand in its way, the research endeavor becomes self-defeating (at least in moral terms) and the researcher becomes the servant of morally illicit and not infrequently vested self-interest. Viewed from this angle, industrialization and commercialization of scientific research render being a virtuous investigator a much harder goal to accomplish. No doubt, (1) integrity and moral character, combined with the (2) virtues along with (3) the discerning power of prudence, often come up against impediments and barriers raised by forces and mechanisms operational within the commercialized research milieu. This is why the need for ethical and moral formation of physicians, nurses, investigators, and indeed all persons involved with medical and scientific endeavor turns out to be so urgent, perhaps more urgent than ever.

3.2.3 Ruth Faden, Tom Beauchamp, and James F. Childress

The analysis of these thinkers revolves around two influential books: *The Principles of Biomedical Ethics* by Tom L. Beauchamp and James F. Childress, further referred to as *Principles*, and *A History and Theory of Informed Consent* by Ruth R. Faden and Tom L. Beauchamp, which will be referenced as *History and Theory.* I have already examined the historical part of the latter book earlier in this study.

Beauchamp and Childress, both extraordinary figures in American bioethics, are commonly associated with principlism and regarded as its canonical representatives. Principlism can be defined as a methodological approach in ethics featuring four prima facie principles that comprise "the proper justificatory framework for bioethics," namely: the principle of respect for autonomy, nonmaleficence, beneficence, and justice (Ainslie 2004, 2100). In this context, with its goal set on the protection of autonomous choices (Beauchamp and Childress 2001, 77), informed consent functions as rooted in and expressive of the principle of respect for autonomy[64] in which it finds its moral justification (Faden, Beauchamp, and King 1986, 294). In both *Principles* and *History and Theory* two types of informed consent are identified: "an individual's *autonomous authorization* of a medical intervention or of participation in research" (Beauchamp and Childress 2001, 78) (sense_1) and "a legally or institutionally *effective* . . . authorization from a patient or a subject" (Faden, Beauchamp, and King 1986, 280), which is "analyzable in terms of rules governing informed consent in public policy and institutional contexts" (Faden, Beauchamp, and King 1986, 274) (sense_2).

Satisfying sense_2 informed consent, which occurs by compliance with research-ethics rules and various policy regulations, does not automatically ensure that all the requirements of sense_1 consent are met. Ethics codes and policy guidelines dealing with experimentation on human subjects and implementing to a great extent the sense_2 approach to consent usually fail to take this sufficiently into account.[65] It is not hard to envision situations in which all procedural standards have been met and yet securing of informed consent is little more than getting a signature on a consent form. Such situations still occur too often to be classified as uncommon; unfortunately, they endanger patients or subjects. Obviously, the converse may also happen: "a patient or subject can autonomously authorize an intervention, and so give an informed consent in sense_1, and yet *not effectively authorize* that intervention in sense_2" (Faden, Beauchamp, and King 1986, 281). Ideally, the two senses should be allowed for and included in the definition of informed consent, although not on equal terms. According to Faden, Beauchamp, and Childress, although "the conditions of sense_1 are not logically necessary conditions for sense_2 . . . they ought to serve . . . as the benchmark or model against which the moral adequacy of a definition framed for the $sense_2$ purposes is to be evaluated" (Faden, Beauchamp, and King 1986, 284). With this in mind, Beauchamp and Childress offer in *Principles* this outline of the elements of informed consent incorporating both $sense_1$ and $sense_2$:

I. "Threshold Elements (Preconditions)

1. Competence (to understand and decide)
2. Voluntariness (in deciding)

II. Information Elements

3. Disclosure (of material information)
4. Recommendation (of a plan)
5. Understanding (of 3. and 4.)

III. Consent Elements

6. Decision (in favor of a plan)
7. Authorization (of the chosen plan)" (Beauchamp and Childress 2001, 80).

A three-point qualification has been offered by Beauchamp and Childress with respect to this description. First, in order to become as exhaustive as possible, the account of informed consent should allow for informed refusal. Therefore, wherever they "use the phrase informed consent", the authors "include the possibility of informed refusal." Second, "consent for research does not necessarily involve a recommendation. If a recommendation is made, it may be quite different from recommendations in clinical medicine." Third, competence counts as a preliminary supposition conditioning the possibility of informed consent rather than sensu stricto its element (Beauchamp and Childress 2001, 80).

Throughout all the threshold, information, and consent elements, both $sense_1$ and $sense_2$ of informed consent pervade the above definition, although not in the same measure. In general, threshold and information elements in the main embrace $sense_2$, whereas $sense_1$ comes to the fore as far as both decision and authorization are concerned. Still, although at varying levels and to different degrees, the two consent senses meet and operate together. Competence can be employed as an illustration of this union: on the one hand it serves as a gatekeeper in health care "by distinguishing persons whose decision should be solicited or accepted from persons whose decisions need not or should not be solicited or accepted" (Beauchamp and Childress 2001, 69); thus competence firmly connects with $sense_2$. Yet, on the other hand, it is predicated on autonomy ($sense_1$) because a substantial deficiency of autonomy typically results in incompetence.

A strong reliance on autonomy also characterizes the second threshold element, voluntariness, which, if taken broadly, remains undistinguishable from autonomy. In the interest of precision and clarity, our authors constrict the former to signify willing "the action without being under the control of another's influence" (Beauchamp and Childress 2001, 93). But even in this narrower sense voluntariness is not divorced from autonomy; in fact, it can be satisfied only when the criteria for substantial autonomy are met. In other words, from among many influences, some are controlling (depriving of autonomy) and some are not. Both *Principles* and *History and Theory* particularize this by pointing to three types of influence: coercion, manipulation, and persuasion. "Coercion occurs if and only if one person intentionally uses a credible and severe threat of harm or force to control another," whereas "in *persuasion* a person must come to believe in something through the merit of reasons another person advances" (Beauchamp and Childress 2001, 94). It follows that coercion always robs an act of autonomy and thus invalidates informed consent, while persuasion never does. As a matter of fact, persuasion, defined by means of appeal to reason, not only does not endanger patients' or subjects' autonomy but also supports and facilitates their consent. Still, between persuasion and coercion much space exists for numerous forms of manipulation, understood as an influence that is neither persuasive nor coercive. Whatever its nature, manipulation, as long as can be resisted, does not control a person and remains congruent with autonomous decision making. Since it is not *full* but *substantial* autonomy that is a prerequisite of autonomous choice, manipulation, by circumscribing autonomy, still does not invalidate autonomy.[66]

Having said this about the preconditions of informed consent, I turn attention now to its elements as laid out by our authors. The first of these is disclosure, whose "conditions for informed consent are central to the history of informed consent in sense$_2$, because disclosure has traditionally been a *necessary* condition of effective informed consent (and sometimes a sufficient condition!)" (Faden, Beauchamp, and King 1986, 281). Our authors disapprove of this priority treatment of disclosure since a person may have sufficient understanding of information and thus give a valid and substantially autonomous sense$_1$ informed consent without any information having been disclosed. An example might be a medical doctor consenting to a surgery to be performed by another physician. Nevertheless, in most cases, in order to provide "an adequate basis for decision making . . . professionals are . . . obligated to disclose a core set of information including (1) those facts or descriptions that patients or subjects usually consider material in deciding whether to refuse or consent to the proposed intervention or research, (2) information the professional believes to be material, (3) the professional's recommendation, (4) the purpose of seeking consent and (5) the nature and limits of consent as an act of authorization" (Beauchamp and Childress 2001, 81).

In order to comprehend fully and to evaluate these requirements of the professional's recommendation one needs to perceive the meaning of material information. First of all, disclosure does not imply divulging *all* relevant information. To endeavor to do that would be neither possible nor advisable since it would likely result in informational overload and thus overburden a patient or subject. In other words, overdisclosure, seen by our authors to be potentially as problematic as underdisclosure, not only does not help but also can seriously impair understanding, which is to say, can compromise the quality of autonomous choice (Faden, Beauchamp, and King 1986, 306–308). Thus, from a set of relevant information, a subset of important, that is, material information should be selected. The crucial qualification is that material information does not pertain to all the facts and details that seem pertinent in light of a certain type of objective standard, but rather to those that prove germane for a concrete subject (subjective approach). Consequently, the same piece of information may turn out to be material for one person but may be totally non-material for another. In order to glean what information actually matters, every effort ought to be made on the part of the professional to set up effective communication with a patient or subject.

The communication process, another word for dialogue, proves necessary for understanding, which is the third informational element of informed consent (Faden, Beauchamp, and King 1986, 308). In fact, a close link between disclosure and understanding exists, as the former is meant to result in the latter. A somewhat related connection also occurs between disclosure and recommendation, which is the second informational element in the definition of informed consent. As shown above, however, the list of requirements for disclosure already includes "the professional's recommendation." Logically, one may then ask why our authors still treat recommendation separately as they enumerate the elements of informed consent.

By contrast, the separate status of understanding seems perfectly justified. Not only is understanding conceptually independent (as it deals more with processing of information than mere disclosure) but it also plays a different role in the process of obtaining informed consent. Faden and Beauchamp describe three categories of understanding: *how, that,* and *what.* While the first relates to competence (one understands *how* to do something), the last two reveal close ties to consent elements (decision and authorization). As our authors explain, patients or subjects "come to understand *that* they must consent or refuse a particular proposal by understanding *what* is communicated in an informational exchange with a professional" (Faden, Beauchamp, and King 1986, 250).

Thus, as disclosure conditions understanding, understanding directly affects the extent to which the criterion of autonomous decision and authorization is met. To secure autonomous consent, *substantial* rather than *full* understanding needs to be ensured. "The condition of substantial understanding demands apprehension of all the *material* or *important* descriptions—but not all the *relevant* (and certainly not all *possible*) descriptions" (Faden, Beauchamp, and King 1986, 302). Comprehension of all the *relevant* or *possible* descriptions would amount to full understanding. This would be too demanding and hence impractical and not a serviceable requirement to apply universally as a necessary threshold for informed consent. As Beauchamp and Childress phrase it, "from the fact that actions are never *fully* informed, voluntary or autonomous, it does not follow that they are never *adequately* informed, voluntary or autonomous" (Beauchamp and Childress 2001, 89).

Although indispensable for authorization, substantial understanding may nevertheless prove insufficient to obtain adequately informed consent. For example, having understood all the material information, a patient or subject may still not *accept* it as true. In fact, "a person must do more than express agreement or comply with a proposal. He or she must *authorize* something through an act of informed and voluntary consent" (Beauchamp and Childress 2001, 78). Authorization can be described as permission given by a patient or research subject to do something to him or her. Put differently, understanding by someone of *what* they are authorizing, although imperative, does not suffice. The understanding of the fact *that* they are authorizing needs to be attained as well.

To sum up, I regard the account of informed consent by Faden, Beauchamp, and Childress one of the most comprehensive ever offered in professional literature as far as its non-historical and non-philosophical content is concerned. The interdisciplinary approach combining historical, philosophical, medical, psychological, and social intricacies of informed consent, characteristic especially of *History and Theory,* deserves appreciation. Likewise, numerous distinctions—arguably the most critical being that between $sense_1$ and $sense_2$ informed consent—along with many case-based illustrations of concrete ethical dilemmas not only reveal a good knowledge of clinical and research realities but also help to avoid a number of the traps and confusions found in the accounts of some other scholars.

Having said this, however, I must note a number of inadequacies and oversimplified conclusions in the historical part of our authors' account, as outlined in the first chapter of this study. Moreover, the philosophical foundations on which Beauchamp and Childress construct their interpretation of informed consent are seriously flawed. While space limitations prohibit a more detailed critique, several brief comments still ought to be made. In general, the most basic problem with the four principles is their prima facie standing.[67] As some commentators have pointed out, not only are the principles "not justified by means of single theory," but they also "offer no real guidance in cases where the principles clash." (Ainslie 2004, 2102; see also Pellegrino and Thomasma 1993, 58–60, 187–95). Nor does the balancing method of reflective equilibrium in the form developed by John Rawls in his *Theory of Justice*, which is adopted by our authors, bring much help.[68] By the same token, the question of why only these and no other principles have been selected remains unresolved (Ainslie 2004, 2103). Naturally, these deficiencies gravely affect our authors' account of informed consent as entirely rooted in and derivative of the prima facie principle of respect for autonomy.

In addition to its fundamental historical and philosophical shortcomings, our authors' treatment of informed consent, although it contains scores of valuable distinctions, nonetheless silently passes over at least three factors I consider crucial to the process of informed consent.

First, despite the attention given to effective communication and dialogue in the process of obtaining consent, no connection has been demonstrated between virtue and moral character on the one hand, and informed consent on the other. Although, aside from the context of informed consent, Beauchamp and Childress touch upon virtue, their approach is far from precise and exhaustive, as shown by Robert Plich (Plich 2006, 224–36). Not only do our authors confuse the classic concept of virtue with its modern successors but they also quite arbitrarily pick out five virtues, which they contend are essential to biomedical ethics, but which on closer inspection turn out to derive from the four principles. This raises fundamental questions about the accuracy of the methodology used by Beauchamp and Childress.[69] What is more, the cardinal virtues, including prudence, which authorities, including Ramsey, Pellegrino and Thomasma have shown to be critical to research and informed consent, have been wholly dismissed with just one statement according to which the idea of cardinal virtues "has never been made very clear in moral writings" (Beauchamp and Childress 2001, 32). This statement, far from clear in itself, becomes even more surprising when one realizes that no further explanation or reference has been offered to support it.

Second, although our authors allow for counseling in the informed-consent process, they understand it narrowly and do not provide satisfactorily for conscience or for its formation. As is the case with virtues, the notion of conscience presented in *Principles* is far from clear and consistent. In effect, otherwise useful statements regarding, for example, persuasion, manipulation, and coercion, are largely unworkable unless employed under the guidance of prudence and a well-formed conscience. The authors realize that the boundaries between these types of influence are not always sharply outlined and hence are hard to distinguish. Yet, the only tool they offer to cope with this issue is a subjective standard of making the discernment entirely conditional on individualized resistance on the part of patients and subjects to the influence which, aside from other factors, depends to a great extent on the way the influence was perceived. This standard, however, is badly defective because human perception is changeable, all the more so when patients are involved, because of the effects of therapy and medication. Under such circumstances, with no sound concepts of virtue and conscience at hand, the approach endorsed by Beauchamp and Childress enhances intuitive and hence, even if not necessarily always irrational, nonetheless anti-rational solutions.

Third, the concept of integrity (Beauchamp and Childress 2001, 35–36), although not foreign to our authors, totally escapes their attention in their account of informed consent. Given the previously outline analysis by Pellegrino and Thomasma, which points to the primacy of integrity over autonomy, this omission by Beauchamp and Childress cannot be condoned.

3.2.4 Germain Grisez

Germain Grisez's intellectual bent, in contrast to that of the authors treated above, proves primarily theological. This is the case despite its appreciation for and not infrequent recourse to philosophy and it obviously influences his analyses of informed consent. Overall, although not reluctant to admit that "the informed consent requirement is not a moral absolute, but only a general duty open to exceptions" (Grisez 1997b, 280), Grisez draws attention to the fact that failure to secure valid consent from competent patients severely violates "their fundamental and legal rights: both their right to decide what potentially burdensome things will be done to them and their right to shape their health care so that it will be harmonious with other aspects of their lives." He further claims that the most important of these other aspects is faith in God, which sheds supernatural light on moral truth, with respect to which every human bears certain responsibilities. One of these is the responsibility for one's own life and health as entrusted to each by God. Therefore, patients whose informed consent has been neglected "are not helped to understand their problems and assume responsibility for them as they must if they are to care for themselves." In truth, it is not only the lack of informed consent that strips those subjected to medical care of their rights and prevents them from taking real responsibility for their health, but also any reduction of informed consent to pure formality that curtails the entire process to merely getting the patient's signature on the consent form (Grisez 1997b, 347).

Echoing Pellegrino and Thomasma, Grisez rejects the idea that informed consent can be anchored in autonomy perceived as a patient's mere "capacity to determine what is good for himself or herself" because "autonomy, so understood, is an instance of the error of subjectivism." Instead, the obligation to secure informed consent "is grounded in the facts that only they can mediate among their own interests, integrate medical treatment with other aspects of their lives, limit it in accord with their other responsibilities, and effectively carry out their own part in it." It follows that "the consent should not be a blank check, licensing care providers to do whatever they consider appropriate, but should be informed and specific" (Grisez 1997b, 330).

Even though Grisez in his writings does not dwell at length on informed consent in terms of its nature, elements, and possible applications, his approach needs to be appreciated and considered relevant on account of two important analyses which it features: first, the analysis of significant risk, a topic critical to informed consent in general and proxy consent in particular (see May 2007); second, the account of prudence and conscience, the importance of which cannot be overestimated in assuring ethical conduct of research involving humans as subjects.

Regarding the first feature, Grisez offers a lucid description of significant risk as "any risk beyond the level of life's common risks" (Grisez 1993, 534). Commenting on this in the context of parents consenting to nontherapeutic experimentation on their children, William E. May notes that "parents frequently take their children, including babies, on automobile trips not undertaken for their benefit (e.g., to purchase clothes for the mother), and such journeys surely involve some risks, but . . . risks of this kind are acceptable both for oneself and for those for whom one cares" (May 2007, 245). In other words, by consenting to nontherapeutic experimentation for their children, parents act in a morally justifiable way as long as no significant risk is involved.

As described in the earlier parts of this work, the risk-benefit assessment is among the core issues marking the informed-consent process. Often the question about the amount of risk involved proves critical to the patient's or subject's final decision regarding participation. Whenever doubts exist as to whether or not the risks involved in a given procedure remain at the minimum level or already extend beyond it, the standard of life's common risk can serve as a criterion in arriving at the answer. Notably, Grisez's understanding of minimal risk relates closely to Pope John Paul II's depiction, invoked above and employed by the *Charter for Health Care Workers*, according to which the risk present in nontherapeutic experimentation should not exceed the risk of any non-substantial physical impairment.

Regarding the second feature, Grisez calls prudence “moral wisdom” and situates it in the context of seeking moral truth. The reason why moral truth ought to be sought, according to Grisez, pertains to its universal and ultimate character: it guides people’s choices and shapes their lives. Consequently, everyone ought to strive to find it. Because of the fallen condition of human nature, which, however, has also been redeemed by Christ and called to eternal glory, moral truth cannot be wholly grasped at the merely natural level. Indeed, its comprehension needs to be buttressed and complemented by the Gospel so that it might become fully capable of directing Christian life. As one seeks for moral truth and fulfills his or her responsibilities toward it, one’s conscience is formed and one’s prudence develops. Although other virtues are also acquired during this process, prudence proves most central because it aids and eases moral reasoning. In other words, even if one knows moral truth and looks to it for guidance, it still takes prudence to apply this knowledge successfully in a situation entailing moral choice. It is precisely this application that Grisez names practical reasoning. The guiding and facilitating role that prudence plays allows the acquisition of proficiency in choosing rightly, so that over time “it becomes second nature for a thoroughly prudent person” (Grisez 1993, 246). To put it otherwise, genuinely prudent individuals “spontaneously fulfill their responsibilities with respect to moral truth” (Grisez 1993, 247).

Seeking moral truth, as Grisez points out, is equivalent to forming one’s conscience.[70] Conscience, in turn, signifies “a person’s apprehension of moral truth considered insofar as it serves to direct particular choices or evaluate them retrospectively; in other words, it is one’s last, best judgment of the moral truth about a choice one is about to make or has made in the past” (Grisez 1993, 249). It follows that the better formed their conscience is, the better humans know moral truth, and vice versa. While conscience provides for moral truth, it is the virtue of prudence that applies it in particular cases, as was inferred from the passage from the *Catechism of the Catholic Church* which was invoked above when examining Paul Ramsey’s account of informed consent.

Undoubtedly, Grisez’s analyses of conscience and prudence supplement not only Ramsey’s presentation but also that of Pellegrino and Thomasma. In spite of the fact that the latter-named authors highlight the relevance of truth construed as the good which is internal to the practice of research as well as the principal value of science, they elaborate neither on theological implications thereof nor on its connections with conscience. Much the same can be asserted with respect to their discussion of prudence, which they recognized yet examined only on philosophical grounds. Still, even within the philosophical framework, the link between prudence and conscience has been overlooked and no serious analysis of the latter has been provided. On that account, Grisez’s approach, even if not conversant with the intricacies of medical procedures, provides more comprehensive grounds for thorough inspection of the realities of research with human subjects in general and the requirement of informed consent in particular as far as moral truth, conscience, and prudence are concerned.

3.3 Exceptions to Free and Informed Consent

With the exception of the CIOMS Guidelines, the human-rights documents surveyed in the first chapter of this study say little about exceptions to informed consent. As a matter of fact, the Nuremberg Code does not provide for any exceptions whatsoever, nor does the Belmont Report. Still, in contrast to the former, the latter regards proxy consent as appropriate in the case of incompetent subjects. The Declaration of Helsinki, in turn, from its second version dated October 1975 through its fifth version issued in 1996, stipulates in the second part dedicated to medical research combined with professional care, that "if the doctor considers it essential not to obtain informed consent, the specific reasons for this proposal should be stated in the experimental protocol for transmission to the independent committee" (n. II.5).

Regrettably, the exception provision thus articulated in Helsinki, as already noted earlier in this study, does not specify what sorts of departures from informed consent are legitimate nor does it provide for the requirements to be met by the potential members of "the independent committee." Additionally, the way Helsinki II–V chose to put the matter invites ambiguity with regard to experimental treatment: is it *any* experimental treatment or only the ones that are part of a research project that ought to be reported to the independent committee?[71] Furthermore, can both competent and incompetent patients be dispensed from informed consent whenever the physician finds it justifiable? Or perhaps only the incompetent can be excluded from what otherwise constitutes a mandatory requirement. The absence of such words as "competence," "emergency," "waiver," and "exception" not only complicates the potential search for the answer to these questions but also encourages an unrestrained or even arbitrary reading of the Declaration's provision.

Perhaps mindful of this problem, Helsinki VI clarifies its exception clause so as to concern only incompetent subjects who are to be involved in research (n. 26). Nevertheless, by placing the exception provision not in the section dedicated to research combined with professional care as was the case in the previous versions of the Declaration, but in the segment dealing with basic principles for all medical research, Helsinki VI suggests the justification of the employment of a subject who has not given consent in nontherapeutic research as well. Yet, no further specification has been given with regard to the nature and scope of such an exception.

On the other hand, however, Helsinki VI sets forth two additional safety provisions regarding subjects. First, "research on individuals from whom it is not possible to obtain consent, including proxy or advance consent, should be done *only if* the physical/mental condition that prevents obtaining informed consent is a necessary characteristic of the research population" (emphasis added). Second, "the specific reasons for involving research subjects with a condition that renders them unable to give informed consent should be stated in the experimental protocol for consideration and approval of the review committee. The protocol should state that consent to remain in the research should be obtained as soon as possible from the individual or a legally authorized surrogate" (art. 26). In summary, although it has nothing to say about the nature and scope of possible exceptions, the Declaration of Helsinki hints that even if the initiation of research in some instances can be found permissible without prior informed consent, the consent still needs to be secured as soon as possible to make it legitimate for the investigator to continue research. In other words, by requiring what has been referred to by professional literature as "deferred consent" (see e.g. Brody 1998, 39; Levine 1995; Morris 2005), *Helsinki* perceives the consent exception as conditional and temporary. Still, nothing has been said to indicate the instances in which initiation of research without consent may be approved.

Notably more specific than *Helsinki* in this regard are the CIOMS Guidelines, which not only uphold the notion of deferred consent but also take the time to address and allow for emergency use of an investigational therapy which the Guidelines, in contrast to *Helsinki*, distinguish from investigational interventions that are part of the research project (see commentary to Guideline 2). The Guidelines also state that "waiver of informed consent is to be regarded as uncommon and exceptional, and must in all cases be approved by an ethical review committee" (Guideline 4). As already noted in the first chapter, the ethical committee is allowed to authorize such a waiver only when minimal risk is at stake and when an attempt to secure consent would render the research infeasible. The Guidelines recognize that a consent exception may be excusable in case of emergency, when "the investigation cannot be done with patients who can give informed consent in time and there may not be time to locate a person having the authority to give permission." Yet, even if this is the case, "the researcher must secure the review and approval of an ethical review committee before initiating the study" (commentary on Guideline 6). Interestingly, while envisioning certain consent exceptions, the Guidelines prohibit any waiver to compensation for participation in research (Guideline 19). Both the CIOMS Guidelines, by far most articulate among the human-rights documents in providing for exceptions from informed consent, and the Declaration of Helsinki with its vague and ambiguous wording, equip ethical review committees with the sole and decisive power to authorize consent exception in research. Despite that, however, instructions offered by both documents are limited, open to varying interpretations, and far from concrete. These shortcomings no doubt contribute to substantial differences in ethical committees' verdicts regarding research projects and the safety of subjects that they employ.

Another issue relates to the terminilogy used in reference to exceptions. For example, the language used by the CIOMS Guidelines does not respond sufficiently to the difference between exception and waiver recognized by numerous contemporary bioethicists. In fact, the CIOMS document treats the two as if they were interchangeable.[72] Furthermore, the lack of precision of the CIOMS and Helsinki language results in the absence of tangible directives regarding therapeutic privilege. It is far from clear to what extent both documents support the employment and application of tangible directives in research. The following discussion addresses this issue of terminology.

Exception in a strict sense, waiver, and therapeutic privilege constitute what is referred to as exception (in a general sense) from informed consent.[73] These three instances need to be distinguished from one another and systematized under the broad category of exception in order to minimize ambiguity and provide more precise and reliable tools for the decision-making process.

Exceptions in a strict sense can take place with regard to both competent and incompetent persons. The former case is rare and, strictly speaking, pertains to informed refusal rather than informed consent. In other words, if a competent person refuses treatment, the refusal can sometimes be overridden and this is likely to occur in two cases: first in the case of pregnant women, and second with respect to parents, usually mothers, of minors.[74, 75] While the former exception, although uncommon, may still be observed in legal practice, the latter belongs almost entirely to the past. As far as the former is concerned, it is sometimes believed that pregnant women cannot refuse life-sustaining treatment if the fetus can be rescued. Yet given its highly contentious character, this belief has not always been shared by courts with their decisions that allowed both the woman and the fetus to die if the treatment was denied. When it comes to parents and their minor children, "courts have sometimes required reasonable treatment to save the lives of parents whose children would otherwise be left orphans, but they have recently stopped doing this ... as American law has been moving further in the direction of the defense of the rights of individuals, and of each person's right to control his or her body regardless of other factors" (Kelly 2004, 145).

Exceptions to informed consent with respect to incompetent patients happen mostly "in emergencies when ... no surrogate is available and the need for treatment is urgent" (O'Rourke 2000, 200).[76] Under such circumstances the consent is presumed (see e.g. May 2003b, 180), that is, "a physician judges that, if the patient could be consulted, he would give his consent" (Kelly 1958, 38; O'Donnell 1996, 52).[77] Whereas no credible presumption exists that would justify exceptions from informed consent in research on competent subjects, consent can be legitimately presumed in emergency clinical trials involving incompetent patients, suffering, for instance, from such conditions as cardiac arrest, profound stroke, or trauma (Sugarman 2007, 12). The section called the "Final Rule", Part 50.24 of Title 21 of the Code of the Federal Regulations[78] (21 CFR 50.24), issued in 1996 by the Food and Drug Administration (FDA) along with the Department of Health and Human Services, mandates that Independent Review Boards (IRBs) may approve exceptions from informed consent when (1) "human subjects are in a life-threatening situation, available treatments are unproven or unsatisfactory, and the collection of valid scientific evidence is necessary to determine the safety and effectiveness of a particular intervention." In addition, a number of other conditions must exist, the most important of them being (2) infeasibility of securing informed consent and (3) "the prospect of direct benefit to the research subject." Moreover, (4) it needs to be clear that "the research study could not be practicably carried out without the waiver" and finally, (5) "attempts to contact the research subject's legally authorized representative within a defined therapeutic window of time must be made."

In contrast to consent exceptions that, except on fairly rare occasions, do not engage competent patients, consent waivers imply competence. According to Beauchamp and Childress, "in the exercise of a waiver, a patient voluntarily relinquishes the right to an informed consent and relieves the physician from the obligation to obtain informed consent. The patient delegates decision-making authority to the physician—or to someone else—or asks not to be informed. In effect, the patient makes a decision not to make an informed decision" (Beauchamp and Childress 2001, 92). Jessica Berg et al. note, by way of summary, that basically patients "should be able to give up the right to information without relinquishing the right to decide, or vice versa. Alternatively, they might determine that they wish neither to be informed nor to decide" (Berg et al. 2001, 88). David Kelly adds that "this is most likely to happen when patients feel that someone else ... will understand the medical complexities better than they could themselves" (Kelly 2004, 147). Still, the legitimacy of waiver comes into question when the health-care interests of other people are involved. Michael Brannigan and Judith Boss warn that when "a patient's waiver of the right to know may jeopardize others, the waiver is generally not justifiable."[79]

Therapeutic privilege resembles waiver in that it also implies competence. Yet, while exceptions and waivers are generally accepted and are considered relatively noncontroversial, the therapeutic privilege, although recognized by U.S. courts and medical authorities (Engelhardt 1996, 317; see also Berg et al. 2001, 80–82; Brannigan and Boss 2001, 126–28; Van den Heever 2005, 420; Sirotin and Lo 2006, 314), stirs up a great deal of emotion, often vented in heated debates. Overall, "it permits the doctor to withhold information when, in his or her judgment, disclosure would be detrimental to the patient's interests or well-being" (Levine 2003a, 201). Faden and Beauchamp qualify that the information at stake needs to be material (Faden, Beauchamp, and King 1986, 37). Still, "the precise formulation of this therapeutic privilege varies across legal jurisdictions. Some formulations permit physicians to withhold information if disclosure would cause *any* countertherapeutic deterioration in the patient's condition. Other formulations permit the physicians to withhold information if and only if the patient's knowledge of the information would have serious health-related consequences." In its narrowest sense, "the therapeutic privilege can be validly invoked only if the physician has sufficient reason to believe that disclosure would render the patient incompetent to consent or to refuse the treatment" (Beauchamp and Childress 2001, 84).[80] In such a situation, as noted by H. Tristram Engelhardt, therapeutic privilege can be construed as a type of emergency (Engelhardt 1996, 317). Generally in an emergency, physicians proceed immediately with treatment according to professional standards and without informed consent, which is justifiable on account of the seriousness of the patient's condition and the impossibility of consulting a proxy. Although ordinarily not as urgent and compelling as classic emergency cases, circumstances warranting therapeutic privilege are similar in that a patient cannot be consulted in view of solid evidence of his or her vulnerability. The condition of the patient is such that, if the required information were passed along, the patient would be under such stress that a competent decision would be impossible. Consequently, if consultation with the patient regarding some aspects of his or her condition is not advisable, the patient's consent can be presumed (Engelhardt 1996, 318).

On the other hand, as reported in 1982 by the Presidential Commission for the Study of Ethical Problems in Medicine, "there is much to suggest that therapeutic privilege has been vastly overused as an excuse for not informing patients of facts they are entitled to know" (President's Commission for the Study of Ethical Problems ... 1982, 92; quoted in Berg et al. 2001, 80). It appears that at least some physicians may still be more likely to give up too quickly on providing information to the patient. Because often it is not so much the character of the information disclosed but the way it is disclosed that upsets the patient, it seems clear that recourse to therapeutic privilege could have been avoided at least in some circumstances, had the physicians taken the time to communicate what is material to the patient in a gentle and respectful manner. Moreover, given contemporary clinical settings, withholding information from the patient frequently initiates a chain-reaction necessitating withholding more information which in turn either implies or adds up to or leads to deception (see Sirotin and Lo 2006, 314). Interestingly, current studies provide solid evidence that patients in most cases actually want to know the truth and often more harm can be caused when the information is kept from them than when it is passed along at an appropriate time and in a tactful way (Berg et al. 2001, 84–85).

Canterbury vs. Spence, the watershed case for both informed consent and therapeutic privilege, points out the danger "that the physician may remain silent simply because divulgence might prompt the patient to forego therapy the physician feels the patient really needs." Although not explicitly addressed by the court, the temptation to remain silent may obviously take a heavy toll also in therapeutic-research milieus. In *Canterbury*'s view, it is chiefly the information about the risks involved in the treatment to be administered that can remain undisclosed, whereas all other information ought in principle to be made known to the patient. Therefore, Berg et al. reason that informed consent, even if based on less than adequate disclosure, should not simply be eliminated when therapeutic privilege is invoked. This conclusion notwithstanding, *Canterbury* suggests that recourse to a family member would still be in order, with the aim of revealing the information withheld from the patient and securing proxy consent for the proposed treatment.

In light of the analysis of the elements of informed consent previously addressed, it is clear that therapeutic privilege directly bears upon the first constituent of the element of information as outlined by Beauchamp and Childress, that is, the disclosure of material information. As has been previously stated, full disclosure of all information is neither possible nor suitable. It follows, then, that fully informed consent is not feasible either, unless by full disclosure one understands a comprehensive revelation of all material information rather than of any relevant or possible information. Hence, not every withholding of information satisfies the criteria for therapeutic privilege. To employ therapeutic privilege means, in a strict sense, to withhold neither irrelevant nor relevant information. Rather, it means not disclosing material information understood as a subset of relevant information. In effect, the question of what does and what does not count as material information turns out to be crucial in determining when therapeutic privilege takes place. Certainly, if a physician resists divulging some relevant information, this does not necessarily entail the therapeutic privilege.

Three competing standards of disclosure have been recognized in current bioethics literature as tools in determining what sort of information counts as material and hence contributes to the therapeutic privilege.

- The first is the standard of professional practice, also called the "reasonable doctor" standard. It maintains that a doctor should "disclose what colleagues similarly situated would have disclosed in similar circumstances" (Veatch 2003, 78). This presumes that "disclosure, like treatment, is a task that belongs to physicians because of their professional expertise and commitment to the patient's welfare" (Beauchamp and Childress 2001, 82). In other words, it is only the practitioner who knows what information counts as material. Likewise, it is only he or she who decides when and how much should be divulged to a patient in a given instance.
- The second standard is that of the "reasonable person," according to which "the physician must disclose what a reasonable patient would want to be told or find significant, even if none of the physician's colleagues would agree." Here, a hypothetical, quite undetermined, reasonable person holds responsibility over what is material and what is not.
- The third is the "subjective standard" with the goal of offering "the patient the information he or she would personally find meaningful" (Veatch 2003, 79). Corresponding to "the specific informational needs of the individual person," the subjective standard more than the others imposes on physicians the duty to come to know what the patient needs to know (Beauchamp and Childress 2001, 83), which implies dialogue and trust.

While the first two standards found their application primarily in legal jurisdictions, the third one prevails in today's mainstream bioethics literature. Obviously, all of these standards have their weaknesses: the first poses a threat of unjustified paternalism, the second is too abstract and hypothetical to prove reliable in the decision-making setting calling for concrete solutions, while the third can easily slip from a subjective into a subjectivistic benchmark.[81] Therefore, in my opinion, all three standards should be balanced against each other and implemented together in order to return what is potentially the most comprehensive determination of what material information is to be divulged to the patient. It is clear that such a determination can be rendered only when the notion of medical relationship is employed, with particular emphasis on the moral characters of both parties and on a level of trust making possible a productive dialogue.

It is noteworthy that the Catholic approach, exemplified, for instance, by the *Charter for Health Care Workers* examined above, works well without resorting to the standards of disclosure just listed and thus avoids the difficulties they pose. The Charter provides objective grounds for justifying therapeutic privilege by adopting the principle of therapeutic trust along with the principle of responsibility for health care. These principles (1) presuppose a dialogic and trust-based relationship between the physician/researcher and the patient/subject and also (2) aim at safeguarding the patient's/subject's inviolable good. Therefore, rather than resorting to merely professional or subjective criteria when utilizing therapeutic privilege, the physician or researcher should take the course of action compatible with the person's good, the recognition and successful application of which calls for, among other things, a virtuous character and a well-formed conscience.

Chapter 4
Introduction to the Issue of Proxy Consent

In brief, an immediate need for proxies arises when patients or research subjects are unable to give their own meaningful and legally valid informed consent because of insufficient or compromised competence or the complete lack thereof. This point of view has been endorsed by all of the human-rights documents examined in the second chapter except for the Nuremberg Code. Accordingly, proxy consent is required unless a consent exception is legitimated and validated by an emergency or other compelling factors as described above.

In a certain sense, proxy consent itself may be viewed as a type of exception from informed consent, although it is hard to find any support for this in legal theory. Yet, this makes perfect sense from an ethical point of view. According to May, proxy consent should be considered a "misnomer, precisely because the potential subjects are noncompetent, 'voiceless' persons incapable of giving consent." Therefore, "the consent in question is the *personal* consent of parents and other guardians" (May 2007, 247), rather than the individual consent of their wards. Consequently, an exception to the ward's consent occurs as his or her legal representative's consent is volunteered.

May's observations conform to a broader issue that exists with regard to the term "proxy consent," calling into question its precision and appropriateness. As a matter of fact, various words are in use for what is often labeled proxy consent. A number of these terms are used almost synonymously, although they are often invoked without consistency or logic. Since relatively few authors take the time to define these terms and to distinguish one from the other, I think it constructive to note key distinctions which will clarify their meanings.

To begin with, two widespread legal notions must be recalled. The first, "durable power of attorney," stands for "authorization to act as agent or attorney for another" ("Attorney, Power Of" 2008).[82] In truth, this is a wide-ranging term used internationally in many realms beyond health care. The second is "legally authorized representative" (LAR), defined somewhat loosely in 45 CFR 46.102c as "an individual or judicial or other body authorized under applicable law to consent on behalf of a prospective subject to the subject's participation in the procedure(s) involved in the research." By offering this general statement CFR left a more precise definition of LAR to the discretion of state laws (Kim and Appelbaum 2006, 471).

G. Mazur, O.P., *Informed Consent, Proxy Consent, and Catholic Bioethics*, Philosophy and Medicine 112, DOI 10.1007/978-94-007-2196-8_4,

In bioethics literature both "durable power of attorney" and "legally authorized representative" are on occasion used in the same sense as surrogate, although other names like "committee" or "conservator" may also apply. In my opinion, this conceptual flexibility does not serve the purposes of ethical analysis because it tends to confuse matters rather than to systematize them. Basically, surrogate, another word for deputy, means "a person appointed to act in place of another" ("Surrogate" 2002), from which it follows that a person having durable power of attorney and an LAR should be regarded as types of surrogates rather than as its equivalents. Indeed, even though treated interchangeably with "proxy" by various authors, the word "surrogate" encompasses a broader meaning. Berg et al. point out, for instance, that "particular kinds of surrogates include a guardian, who is a court-appointed surrogate, and a proxy, who is a surrogate selected by the patient himself" (Berg et al. 2001, 109).

One of the well-known methods available for patients to select proxies is broadly known as advance directives.[83] Typically understood as a part of a larger process called advance care planning, "advance directives are oral or written statements in which people declare their treatment preferences in the event that they lose decision-making capacity" (Fischer, Tulsky, and Arnold 2004, 74). The two commonly recognized types of advance directives are proxy directives and instructional directives.

The former, "often referred to as *durable powers of attorney for healthcare,* allow patients to specify a person or persons to make decisions" (Fischer, Tulsky, and Arnold 2004, 75) as well as to spell out the scope of the mandate of that person or those persons in the decision-making process. Regarding that aspect, "some states specifically limit the proxy's authority to make decisions about certain types of treatments, for example, life-sustaining treatment, experimental treatment, psychosurgery, and sterilization" (Berg et al. 2001, 111). Regardless of how extensive their power ultimately turns out to be, the mere possibility of assigning proxies proves particularly useful when persons oppose having their legal next of kin decide for them in the event of substantial deficiency in or complete loss of competence (Veatch 2003, 106). Regarding the terminology, it is worth noting that durable power of attorney in this context pertains to written documents or oral authorizations rather than to a person acting as a proxy.

In contrast, the latter type of advance directives, that is, instructional directives, frequently spoken of as "living wills," record the patient's "wishes, preferences and values" with respect to future treatment (Fischer, Tulsky, and Arnold 2004, 76). Instructional directives can be divided into substantial and insubstantial according to the weight of the wishes they contain (Veatch 2003, 106).

Both proxy directives and instructional directives, despite their validity and several positive characteristics, nonetheless pose serious problems and are subject to ongoing criticism. One of the most commonly expressed concerns, for instance, is whether or not one's preferences, made known prior to the treatment, remain binding when incompetence occurs. It is hard to anticipate decisions concerning future medical alternatives, which are normally predictable only in part; these decisions may fluctuate over time as the patient's condition changes or new, previously unforeseen, factors arise. Likewise, advance directives, usually written at a general level, even if consistent in themselves, seldom succeed in providing sufficient assistance in determining the optimal course of medical action. As recognized elsewhere, "no matter how specific the document, it is unlikely to capture the circumstances of a real clinical situation exactly" (Fischer, Tulsky, and Arnold 2004, 76).

Mindful of these and other difficulties,[84] Scott Kim and Paul Appelbaum, while elaborating on the research consent of persons suffering from dementia, propose what they call "concurrent proxy directives" as distinguished from advance ones. Concurrent directives are possible because some persons incapable of giving informed consent themselves due to diminished competence may still be capable of appointing a proxy. Thus, "in contrast to the completion of an advanced directive, the appointment of a proxy decision-maker need not be thought of solely in terms of planning for the future but rather as something that can be done at the time a decision making situation arises" (Kim and Appelbaum 2006, 470). Needless to say, concurrent directives captured in this way successfully escape many of the difficulties that inherently trouble their advance counterpart.

The status of a proxy acquired by means of advance or concurrent directives differs from the status of an agent who becomes a surrogate by virtue of law rather than by the prior appointment of a now-incompetent person. Typically, according to current legal regulations, it is family members who are selected as surrogates in the absence of advance directives. Many states have established a hierarchy of potential decision-makers that determines the order in which they will be employed as surrogates in cases when the next of kin (usually a spouse) is not available. Such de facto surrogates, however, are ordinarily not as well-equipped with legal decision-making instruments as are the proxies appointed by means of advance directives. For example, one concrete situation in which a proxy receives considerably more decision-making power than a de facto surrogate has been recognized by New York state, where in order to authorize the withdrawal of artificial hydration and nutrition a proxy is required to have a "reasonable knowledge" that the now-incompetent person for whom the decision is being made would express the same wish, while a de facto surrogate must possess "clear and convincing evidence" of the now-incompetent person's agreement (Kim and Appelbaum 2006, 471).

This makes a compelling case for calling for a clear distinction between surrogate and proxy consent. In response, Ashley and O'Rourke propose to introduce another term encompassing both consent types, that is, vicarious consent. The reason is that "a vicar fulfills a duty for another irrespective of whether the other has authorized it" (Ashley and O'Rourke 1997, 350). When combining this suggestion with what has been stated in previous paragraphs, the following conclusion emerges: vicarious consent incorporates by far the broadest scope of meaning and as such it includes proxy consent, by means of which a decision-maker is appointed by the then-competent person, and surrogate consent, where authorization is not done by advance directive. In research settings, the phrase "legally authorized representative" can be used for both consent types. Moreover, as far as proxy consent is concerned, "durable power of attorney for healthcare" proves an equivalent term. Although I find Ashley and O'Rourke's remarks on vicarious consent useful and hence worth keeping in mind, they nonetheless add little to the more fundamental distinction between surrogate and proxy consent which already involves the very rationale on the basis of which the word "vicarious" has been put forward. Moreover, since in the majority of cases ethical considerations do not hinge so much on the way in which surrogates and proxies are called into being, but rather on the very phenomenon of surrogacy, current informed-consent literature strongly tends to use "proxy" and "surrogate" consent interchangeably. I follow this trend in this study when the distinction between proxy and surrogate remains irrelevant for the clarity and intelligibility of the analyses being conducted.

In the end, no matter which term is used, vicarious/surrogate/proxy consent must still be differentiated from both implicit and presumed consent. "Persons consent implicitly when they actually consent to some general line of action, which implies more detailed permissions, as when a patient consents to surgery without specifying what anesthetic is to be used." By contrast, "consent is presumed when it is highly probable that someone who is not able to give because he or she is absent or unconscious would have given it if present or conscious" (Ashley and O'Rourke 1997, 350), as is the case with valid exceptions to informed consent.

Like informed consent in general, proxy and surrogate consent can be employed in the context of either therapeutic or nontherapeutic research. Introduced first by Helsinki I, the differentiation between two major types of research soon became the subject of animated debate, as was previously shown. Called by some "crucial" and by others "unacceptable" and "illogical" (Schaffner 2004, 2328), the therapeutic-nontherapeutic distinction still causes a good deal of contention.

One of the staunchest opponents of the distinction is Robert Levine, who also belongs among the most prominent figures in research ethics. In charging that the distinction lacks logic, Levine nonetheless makes a logical error himself, as he attempts to illustrate his position based on the example of the Declaration of Helsinki (Levine 2003b, 106–107). In a word, what Levine actually criticizes is the limited and imprecise usage of the therapeutic-nontherapeutic distinction by *Helsinki*, rather than the distinction itself. Furthermore, even though many of Levine's observations deserve merit and attention, the conclusion he draws simply cannot be accepted. To put it simply, it is true that the distinction between therapeutic and nontherapeutic research has always been prone to abuses and vagueness, but this is also precisely the reason why it needs to be retained and refined.

A recent instance of this kind of positive approach can be found, to give but one example, in the article of Marilyn Morris and Robert Nelson in which they underscore that the therapeutic components of research ought to be separated from nontherapeutic components, and demonstrate that some procedures might be therapeutic under one set of circumstances and nontherapeutic under another (Morris and Nelson 2007, 942). Likewise, as is widely known, the same protocols for placebo clinical trials may be both therapeutic and nontherapeutic depending on the group of subjects (see e.g. Ashley, DeBlois, and O'Rourke 2006, 116).

By definition, "*therapeutic research* studies the effects of using diagnostic, prophylactic, or therapeutic methods that depart from standard medical practice but hold out a reasonable expectation of success." In contrast, "*nontherapeutic research* ... is not designed to improve the health of the research subject; rather, it seeks to gain knowledge or develop techniques that may benefit people other than the subject in the future" (Ashley and O'Rourke 1997, 346). The criterion of knowledge, or more accurately, generalizable knowledge, plays a critical role in discriminating between research and what is called innovative therapy or nonvalidated practice/therapy (Schaffner 2004, 2328), which is frequently applied when standard therapies turn out to be unworkable. It is agreed that physicians may have recourse to such nonvalidated practices provided that they directly benefit the patient rather than contribute to generalizable knowledge (Veatch 2003, 143).

Still, in my opinion, if the practice in question happens both to benefit the patient and to contribute to generalizable knowledge (an instance not addressed in Levine's critique), it counts as therapeutic research and ought to be provided with a research protocol and reviewed by a research ethics committee. As summarized by Kenneth Schaffner, "for interventions from which the researcher intends to produce new general knowledge, that represent significant departures from accepted practice, and about which there is reasonable uncertainty regarding consequences, including intended outcomes, it would seem mandatory that the researchers develop a formal research protocol to be assessed by an appropriate institutional review board (IRB)" (Schaffner 2004, 2329).

Levine correctly points to one additional class of medical activities, so-called practice for the benefit of others, which references all the definitions of therapeutic and nontherapeutic research, medical practice and nonvalidated practice. Actually, it differs from medical practice "only in that it is not 'designed solely to enhance the well-being of an individual'" (Levine 2003b, 105). Nevertheless, since it covers such activities as organ donations, vaccinations, and quarantines, practice for the benefits of others coincides with research in that it entails burdens to be undertaken by the patients for the advantage of some third parties, which in turn demands the employment of ethical standards intended for research rather than other activities (Levine 2003b, 105).

Ethical standards governing research that engages humans as subjects recognize the privileged position occupied by informed consent, including its vicarious forms. In fact, the distinction classifying research as therapeutic or nontherapeutic helps to make clearer the intricacies of consent and the challenges arising as it is employed. The issues become even more complex when surrogates are involved, the topic which will be the main focus throughout the following chapters.

Chapter 5
Standards for Proxy Consent in the Therapeutic Situation

As the earlier chapters of this study have shown, research may be classified as therapeutic when it is beneficial to its subjects or as nontherapeutic when it does not bring about any medical advantage to those involved as subjects. With a focus on the therapeutic setting, this chapter outlines the most commonly accepted standards for proxy decision making, while the next chapter analyses these standards and seeks to determine the ultimate reason justifying proxy consent in research according to sound Christian anthropology and moral theory.

It is widely accepted in today's bioethics that, in contrast to the situation of competent individuals who are expected to give informed consent on their own, proxies may be appointed to decide on behalf of incompetent patients or research subjects. Incompetence, if not permanent, does not preclude taking into account any relevant wishes or preferences applicable to the patient's current medical condition that may have been disclosed by the patient when competent. While permissibility of proxy consent in therapeutic research typically raises no concerns among bioethicists, the grounds on which it rests are more controversial. Also, the boundaries within which securing proxy consent in therapeutic research meets moral standards are not always precisely delineated. With these issues in mind, I begin the discussion by presenting below the most commonly accepted ways of understanding proxy decision making in present-day mainstream American bioethics.

5.1 Standards for Proxy Decision Making

A well-established taxonomy features three standards for proxy decision making: (1) advance directives, (2) substituted judgment or its refined form known as the pure autonomy standard, and (3) the best interest standard. Furthermore, it is widely accepted that the three standards relate to each other in order of application. For example, if advance directives cannot be applied, then one should turn to substituted judgment/pure autonomy. When substituted judgment fails, then the best interest principle should be utilized (Brock 2004, 2485). While no serious charge can be leveled against the ordering of standards, at least two reasons, in my opinion,

G. Mazur, O.P., *Informed Consent, Proxy Consent, and Catholic Bioethics*, Philosophy and Medicine 112, DOI 10.1007/978-94-007-2196-8_5,

undermine the advisability of giving advance directives a separate standing, in particular as far as research is concerned.

First, advance directives never represent the actual decision that a now-incompetent patient or research subject would make were he or she competent. The actual decision is made by a surrogate. In fact, a great deal of confusion pervades professional literature in this respect. Certain authors seem slow to acknowledge that advance directives constitute at best a prior rather than a concurrent recommendation of the now-incompetent individual. This recommendation was made from a necessarily constrained perspective as it was an anticipatory prediction of a possible future medical condition with very little, if any, knowledge of what the patient's future feelings and preferences would be, not to mention what medical options would be available at that time. Consequently, by their very nature and regardless of how thoroughly and meticulously drawn up they may be, advance directives call for interpretation and implementation preceded by an honest assessment of the current situation of the patient by the proxy.

Second, as far as research is concerned, advance directives (at least in their instructional form) do not bring much value because most of them address only treatment preferences. Hence, although they may turn out to be useful to some degree in the case of therapeutic research, this cannot hold for nontherapeutic settings.

Based on those two reasons, I propose that advance directives be included either in the substituted judgment or in the pure autonomy standard. In fact, the latter option has already been advocated by Beauchamp and Childress (Beauchamp and Childress 2001, 103). Thus the three standards to be presented below are: substituted judgment, pure autonomy, and best interest.

5.1.1 The Substituted Judgment Standard (SJS)

This standard demands that surrogates make the same decisions on behalf of and for any now-incompetent persons (patients or research subjects) that those persons would make themselves in a given set of circumstances if they were competent. Clearly, the required accordance with the incompetent's wishes, goals, and values renders this standard a subjective one. Berg et al. thus equate its aim with the implementation of "the subjective preferences of the patient about and for whom decisions must be made" (Berg et al. 2001, 114; see also Veatch 2003, 119). In order for this goal to be achieved the surrogate needs to be acquainted with the now-incompetent's world of values to the extent that an adequate judgment can be formulated. According to many contemporary authors, the task becomes easier if the person in question has made appropriate decisions in regard to his or her future treatment, such as those frequently featured in advance directives, before lapsing into incompetence.

Although not entirely inaccurate, this belief tends to neglect the fact that advance directives routinely hinder rather than assist the process of gaining knowledge of the now-incompetent's real preferences, as was shown earlier. Regardless of the exact form it may take, the expression of prior preferences constitutes the condition sine qua non of the SJS's workability. In other words, the standard proves propitious neither for never-competent persons nor for persons who, although once competent, never made their relevant goals and wishes known to their future proxies (Beauchamp and Childress 2001, 100). Even if a once-competent person did not hesitate to disclose his or her goals or wishes, a proxy still needs to make sure that they have not changed in the meantime and that they can be consistently applied in a given medical situation.[85]

In his 1997 article on substituted judgment, Mark Tonelli highlights the fact that the standard itself, not only welcomed but frequently preferred among lawyers, ethicists, and physicians, nonetheless has received different interpretations within each of these professional groups. As the article states, the "clinical approach differs significantly from the standard legal procedure and, less drastically, from the most accepted bioethical recommendations regarding the use of the substituted judgment standard" (Tonelli 1997, 22). Below I present briefly all three approaches to SJS, based, unless stated otherwise, on Tonelli's analyses.

5.1.1.1 Legal Approach

The presence of substituted judgment in law can be traced back to the early nineteenth century, in a case dealing with the management of the estate owned by an incompetent person in England (Tonelli 1997, 23; see also Berg et al. 2001, 114). In the US it was first invoked in 1977 in *Superintendent of Belchertown State School v. Saikewicz* (Peterson 2004, 2924). The *Saikewicz* court stated that the decision concerning the withdrawal of life-sustaining treatment "should be that which would be made by the incompetent person, if that person were competent, but taking into account the present and future incompetency of the individual as one of the factors which would necessarily enter into the decision-making process of the competent person" (Jonsen, Veatch, and Walters 1998, 156). Taking into account the incompetence of the ward, although commendable, does not change the fact that *Saikewicz* nonetheless underestimated the life-long and permanent character of this incompetence. Joseph Saikewicz, "a 67-year-old, severely retarded, institutionalized patient suffering from leukemia" (Jonsen, Veatch, and Walters 1998, 153) never found himself capable of making, much less conveying, a meaningful decision concerning his life-goals, wishes, or preferences. Still the court ruled that the life-sustaining treatment could be abandoned based on six determinants: "Saikewicz's age, the probable side effects of treatment, the low chance of producing remission ..., the certainty that treatment will cause immediate suffering, Saikewicz's inability to cooperate with the treatment," and "the quality of life possible for him even if the treatment does bring about remission" (Jonsen, Veatch, and Walters 1998, 157).

Although one may admit that these factors are sufficient for foregoing life-sustaining medical procedures, one can hardly deny that settling the matter by recourse to SJS begs the question. This is mainly because the court rested its verdict on the court-appointed guardian's opinion regarding what Saikewicz would prefer in a given situation if he were competent rather than on his real preferences which, as noted above, were never revealed. Most commentators admit that the standard utilized in *Saikewicz* is better described as "best interest" (the guardian attempting to identify what works best for the ward based on hypothetical although highly probable "best interests") instead of "substituted judgment" (the guardian endeavoring to identify what works best based on the patient's preferences as articulated when competent). In lieu of Saikewicz's nonexistent judgment, the court came up with its own and assumed that it matched the ward's would-be decision. Interestingly enough, eleven years after *Saikewicz* was decided, Judge Paul Liacos, who authored the decision, admitted that resorting to SJS in this type of situation "involves a legal fiction to some extent." Despite that, however, he still advocated the use of substituted judgment "as preferable to the 'best interest' alternative" (Liacos 1989, 221; quoted in Tonelli 1997, 23).

Tonelli sees the raison d'être for this legal preference as twofold. First, employing SJS rather than any form of an objective standard such as best interest allows the decisions to be discussed in terms of rights. In brief, the assumption is that any competent person has a right to object or to refuse treatment and this right should be extended also to the incompetent. Although such a declaration claims to find support in the U.S. Constitution, it is not evident on what premises this claim actually rests. On the other hand, "this odd assertion of a right of autonomous choice for an incompetent (and therefore non-autonomous) individual provides a degree of protection from the interference of the state that is not produced by any of the objective standards" (Tonelli 1997, 23).

Second, by recourse to SJS, judges can avoid getting involved in the evaluation of the quality of life of the incompetent persons, the inevitable characteristic of any born-in-law objective standard. "If withdrawal or support is ever to be in the best interests of a patient it must be because the quality of that patient's life is below a certain minimum, where death is preferable to continued existence. Such an assessment of quality of life, however, appears to many to be inseparable from a judgment about the value of an individual life."[86] Therefore, courts that mention the quality of life criterion usually do it with great caution and with additional clarification. By contrast, employing SJS under such circumstances makes it possible for competent individuals to quantify their own "personal minimum quality of life" (Tonelli 1997, 23), to be used as a gauge when their competence declines.

Notwithstanding legal preferences for SJS, it is clear that in order to rescue the standard from arbitrariness one must establish evidentiary criteria of patients' previous goals and wishes. This is to say that courts are in need of "an appropriate methodology for evaluating evidence" so that the implementation of SJS might be found defensible on the grounds that it sufficiently reflects the patient's preferences rather than those of a proxy. Unfortunately, the above-mentioned constitutionally protected legal postulate employed, for example, in *Saikewicz* (i.e., that incompetent persons are entitled to enjoy the rights that the competent do, because doing otherwise would amount to discriminating against the incompetent) caused the legal evidence deemed sufficient for invoking substituted judgment to be lowered to an unreasonable level (Buchanan and Brock 1989, 114).

On the other hand, and apart from the medical context, the law generally conditions the quantity of required evidence on the interests of the individual involved in the case: the more basic the interests involved, the more solid the evidence needed. No doubt, when applied in the realm of surrogate medical decision making, this approach helps to answer the question about what quantity and quality of evidence suffices for implementing SJS. For instance, the New York Supreme Court in *In re Eichner* found only "clear and convincing" evidence to be acceptable when the withdrawing or withholding of life-sustaining treatment is at stake. In cases featuring interests less vital than life, perhaps a lower standard of evidence could be used. As the U.S. Supreme Court ruled in *Cruzan*, it is within the discretion of individual states to establish levels of evidentiary standards corresponding to various categories of patients (Tonelli 1997, 24).[87]

5.1.1.2 Ethical Approach

Although, generally speaking, the ethical approach to SJS in present-day secular bioethics is closely related to that observed in the courts, it nonetheless differs from its legal counterpart in that it builds not on the authority of the U.S. Constitution but rather on the acknowledgment of the value of individual autonomy.[88] As Tonelli puts it, "with the recognition of the moral relevance of patient autonomy comes the desire to extend the privileges and protection of autonomy to those who are no longer able to exercise choice in any meaningful way. The claim is that individual autonomy is so vital to medical decision making that it demands to be respected even after it has been irretrievably lost" (Tonelli 1997, 23). This conclusion parallels the legal extension of the right to autonomous choice illustrated above by the arguments used in *Saikewicz*. Even if rejected by certain ethicists like Beauchamp and Childress (Beauchamp and Childress 2001, 99), the autonomy extension is welcomed by others, of whom one influential example is Robert Veatch.

Robert Veatch

When addressing the topic of proxy decision making in his book *The Basics of Bioethics*, Veatch divides incompetent patients into three subgroups: first, formerly competent and now incompetent patients; second, never competent and having family; and third, never competent and without family. Veatch attempts to show how extended autonomy applies to SJS by considering the first two classes of patients. Regarding the formerly competent, the assumption is that a now-incompetent person made a decision applicable to his or her current condition when still competent. It must be noted that making a prior decision of this kind implies something more than merely expressing wishes or preferences, a fact about which Veatch does not seem particularly cautious as he tends to use these terms (decision on the one hand and wish or preference on the other) interchangeably. Nevertheless, he still puts stress on advance directives as a preferred form of decision making before a person falls into incompetency. The principle justifying the implementation of advance directives by the proxy is that of autonomy extended, that is, the perception that the decision, although made in a period of competence, still binds after competence has deteriorated. Analogously to the legal situation presented in the preceding subsection, the broadening of autonomy understood in this way strains the fundamental standards of logic and common-sense convictions. The issue becomes even thornier when a patient loses memory of his or her past self, as often turns out to be the case with those suffering from advanced Alzheimer's disease, as one example. Despite acknowledging this problem, Veatch does not suggest any solution.

As far as never-competent patients with family are concerned, a different type of autonomy extension comes into play. Veatch argues that a certain degree of discretion should be accorded to the family surrogates, typically to the next of kin, even if the course of action they propose deviates to some extent from what would fall under the best interest standard. The deviation can be approved on the strength of the principle of familial autonomy. In other words, the patient's family's autonomy is extended in such a way that it overlays the patient's inferred wishes even if its decisions differ from the options identified as optimal by others. Still, it is agreed that the divergence between the proposition based on autonomy thus broadened and that put forward by an objective standard must not go too far. If it does, it is likely to be questioned by third parties involved (a physician, friends of the patient, etc.) with a possible continuation in the court of law. Such was, for example, the case of Chad Green, who at the age of two and a half was diagnosed with leukemia. The chemotherapy recommended by the physician was denied by the parents on the grounds of its toxic and painful nature. The parents believed that some form of alternative diet could be used instead. In the face of this, the physician sought a court order, which he eventually received and which overrode the parents' resistance and required the administration of the recommended chemotherapy. In the justification of its ruling, the court noted that "the parents had exceeded the limits of reason in pursuing what was in the child's best interest" (Veatch 2003, 117). This example shows that, in some contrast to its individual correlate, familial autonomy in Veatch's view has limits that permit the interference of third parties even to the point of judicial reversal of the family's surrogate decision. It is the recognition of the vulnerability of the incompetent patient which creates the potential for abuse, and that in turn according to Veatch justifies the parties' interference.

Still, one may ask what kind of objective standard should be contrasted with proxy decisions based on familial autonomy and on what basis the latter may be overridden by the former. Following the prevailing legal tendency, Veatch opts for the reasonable-person standard, according to which family's surrogate decisions count as valid unless they extend beyond the boundaries of reason. Even with this qualification, however, Veatch explains neither where the line between the rational and irrational lies nor on what grounds this line ought to be crossed.[89] Likewise, despite arguing in favor of the principle of familial autonomy, Veatch concedes that the rationale on which this principle rests eludes easy identification: "some find it in religious traditions, the divine ordination of the family as a fundamental unit; others in pragmatic consideration such as the probability that the family is more committed to the welfare of the child than any other social group or the need to reward families for their service by allowing some limited discretion." Still, one does not know where Veatch himself finds it. In spite of that, he states that whatever the basis of familial autonomy, "it is widely agreed it exists" (Veatch 2003, 117).[90]

Allen E. Buchanan and Dan W. Brock

Although they do not utilize the principle of familial autonomy, Buchanan and Brock, in *Deciding for Others*, their pioneering book labeled "a leading reference work in the area" (Charland 2008), concur with Veatch on the justifiability of a restricted discretion for family surrogates. Yet unlike him they address this discretion only insofar as formerly competent and now-incompetent individuals are concerned, with the assumption being that during the period of competency, relevant wishes and preferences were disclosed. This position is shared broadly among present-day lawyers, bioethicists, and practitioners. Moreover, Buchanan and Brock emphasize the crucial role that evidence plays in the application of SJS, a factor not considered by Veatch. Since the standard implies "a judgment about what the patient would want if competent, [it] is always to some degree *reconstructive* and *speculative*; it is a hypothetical statement developed from evidence about the patient's past preferences and values. As such, it does not carry the same moral weight as the actual choice of a patient when he or she is competent, nor even the same moral force as a clear and well-documented advance directive" (Buchanan and Brock 1989, 146). Although an advance directive may have a binding status close to that of an actual decision of the competent individual that immediately precedes a treatment or a research procedure, it nonetheless has lesser strength,[91] another fact not clearly conceded by Veatch.

It is precisely because of the limited nature of substituted judgment (that is, the fact that it is never equivalent to the free choice of a competent person) that SJS does not hold an absolutely independent standing. Even family surrogates, although preferable, cannot be an exception to this rule. Buchanan and Brock mention two circumstances under which the substituted judgment of a familial proxy ought to be challenged even to the point of its reversal. First, "if there is a consensus among reasonable and informed persons that the family's construction of what the patient, if competent, would choose is an unreasonable inference from the evidence about his or her former general preferences and values, and if there would be a significant harm or loss to the patient if the family's choice is respected." Second, even if the inference from the available evidence seems reasonable, "it may still be challenged if the decision is likely to result in severe harm to the patient and if the evidence for what the patient would choose is indirect and/or less than fully decisive and consistent" (Buchanan and Brock 1989, 146–47). These two provisions introduce a more protective stand toward an incompetent person than that endorsed by Veatch and based exclusively on the rationality or reasonableness of a familial surrogate's decision. In addition, Veatch's principle of familial autonomy precludes weighing any evidence whatsoever of the patient's wishes as it holds only for permanently incompetent wards.

In Buchanan's and Brock's view, the limited character of SJS, no matter how strong the evidence, can be most clearly captured by referring to the category of basic interests found originally in estate law and noted in the previous subsection. In short, the legal premise is that the basic interests of a now-incompetent should be defended even if doing so conflicts with what the person chose when competent. While in estate law this assertion enjoys absolute force, Buchanan and Brock propose that a more nuanced variant be used in executing substituted judgments under medical conditions. "The variable weight version of the basic interests limitation," as they call it, puts on a sliding scale basic interests and the evidence of the individual's wishes or preferences. When the evidence proves ample, as Buchanan and Brock believe might be the case of advance directives which are clearly thought out, well documented, and unambiguous, the wishes and preferences they contain should prevail. But if no advance directives were issued, "in those cases in which the overall strength of the evidential base is weaker, the 'basic interest' limitation . . . should be given proportionately greater weight." On the other hand, "where the overall strength of the evidential base is greater, the 'basic interest' limitation should be accorded less weight" (Buchanan and Brock 1989, 119–20).

While no doubt plausible, the balancing strategy thus construed nonetheless is open to ambiguity and abuse so long as one does not specify what is actually understood under the basic-interest restriction. Similarly, although relying heavily on the evidence, Buchanan and Brock do not explain how this evidence is to be determined, and the several rule-of-thumb insights on the matter that they offer do not provide much guidance. Last but not least, even though they are more realistic and nuanced in their appraisal of advance directives than is Veatch, Buchanan and Brock still give advance directives priority treatment, a controversial practice in the light of the preponderance of the literature as well as the concise analysis of advance directives offered earlier in this study which reveals serious flaws inherent in both advance directives themselves and in the ways they are employed.

5.1.1.3 Medical Approach

Little has been said in contemporary bioethics literature about the way physicians handle SJS in their day-to-day professional practice. Arguably, the most significant difference between the approach of physicians on the one hand and that characteristic of lawyers and ethicists on the other comes down to the conviction that SJS per se cannot work satisfactorily, but only as paired with an objective criterion. This is best observable in the way medical practitioners deal with the available evidence. Tonelli believes that "when a surrogate communicates to the medical staff a choice seemingly based on substituted judgment, the physician's consideration of that decision starts . . . not with an evaluation of the evidence available to support it, but with a comparison to the course of action that seems to be dictated by the objective standards" (Tonelli 1997, 27). The objective standards as perceived by physicians ordinarily point to what seems indicated from the medical standpoint. Hence, the more a proxy's decisions diverge from medical recommendations, the less the physician is obliged to follow them. This issue often arises without even questioning the reliability of the evidence provided in a given case. For the most part, however, the evidence of the incompetent's prior wishes is not dismissed altogether and comes into play as a second step after an initial comparison to what is medically indicated or, more broadly, to what lies in the patient's best interest, has been made.

These two factors, the evidence and the objective criterion, make up two variables of a sliding scale that determines the physician's reasoning. It follows that little or no evidence is required when the proxy's substituted judgment coincides with the medical indication. In contrast, when the former departs greatly from the latter, the reliability level of desired evidence will be set high. In other words, the bar of required evidence is raised in direct proportion to the degree of disagreement between the proxy's proposed choice of action and that predicated on the objective standard.

Although resembling that of Buchanan and Brock, the physicians' sliding scale strategy differs in that it does not demand for each decision an aggregation of all the available pieces of evidence together with an assessment of their validity and strength. Moreover, the objective standard taken into account by physicians does not necessarily equate to basic interests because a broader criterion of best interests can be invoked as well. Overall, even though both the Buchanan-Brock approach and that of physicians may converge in the results they yield, the latter seems by far more practical and easier to use (Tonelli 1997, 28).

5.1.2 The Pure Autonomy Standard (PAS)

While a number of thinkers such as Tonelli argue that SJS constitutes a preferred method of surrogate decision making, others renounce it on the grounds that when properly applied SJS slips into the pure autonomy standard. As prominent representatives of this type of critique, Beauchamp and Childress emphasize that "an incompetent person cannot literally be said to have the right to make medical decisions if that right can only be exercised by other competent persons" (Beauchamp and Childress 2001, 99). Consequently, they avoid the troublesome assertions made initially by *Saikewicz* and subsequently defended by some bioethicists on the basis of so-called extension of autonomy. Rather, Beauchamp and Childress believe that PAS "eliminates the ghostly autonomy found in substituted judgment . . . and applies exclusively to formerly autonomous, now-incompetent patients who expressed a relevant autonomous preference" (Beauchamp and Childress 2001, 100). In a sense, then, PAS can be thought of as a refined SJS, where an autonomous proxy chooses from among available medical options for and on behalf of a now-incompetent individual whose prior autonomous goals, wishes, and decisional patterns are honored and applied in the current situation.

Even though more reliable than SJS, PAS does not, however, circumvent all of the issues. Beauchamp and Childress record three that bear upon the lack of sufficient evidence supporting the standard's implementation. First, it may occur that a surrogate selectively culls from the ward's previously expressed preferences those that coincide with his or her own views, and then employs only those in arriving at the final decision. Such a practice may relate to what has been elsewhere called "projected judgment" (Brannigan and Boss 2001, 488), in which a surrogate projects his or her own values and implements them rather than those of the ward. Second, "the surrogate's findings might also be based on values of the patient that are only distantly relevant to the immediate decision, such as the patient's expressed dislike of hospitals." Third, "a troublesome problem is that surrogates often assume an explicitness in a patient's directive about the future that does not, with sufficient directness, apply to the decision at hand" (Beauchamp and Childress 2001, 101).

Needless to say, these problems affect both SJS and PAS. Or more precisely, they point to a more elementary question of who the proxy is in the moral sense of the word. This conclusion, nonetheless, seems alien to Beauchamp and Childress, as they do not probe below the solely technical surface of encumbrances that may hamper the sound implementation of PAS. But even on this merely observational level no concrete guidance is proposed on how to protect the incompetent against the proxy's bias. It appears that the only safety barrier in Beauchamp's and Childress's account is the patient's former autonomy and the fact that it connects with the decision now to be made by the proxy. No examination of the quality of evidence supporting prior wishes, autonomously expressed, is offered. On one hand our authors propose that PAS stand on its own, but on the other, in the light of the three practical issues they acknowledge, it becomes clear that PAS alone does not suffice. In other words, recourse to another standard, criterion, or value appears unavoidable in cases when the previous autonomous wishes of the patient turn out to conflict with the actual autonomous preferences of the proxy. Surprisingly enough, this uncomplicated assumption escapes Beauchamp's and Childress's attention.

To be fair to them, it must be admitted that when they address the best interest standard, our authors allow for legitimate replacement of autonomously issued advance directives by a now-incompetent patient, yet the example they give refers to proxy rather than instructional directives. In their own words, "if the designated surrogate makes a decision that clearly threatens the patient's best interests, the decision should be overridden unless there is a clearly worded, second document executed by the patient that specifically supports the surrogate's decision" (Beauchamp and Childress 2001, 102). In other words, even if a proxy directive can be overridden on protective grounds, an autonomously issued instructional directive, even if contrary to the incompetent's best interests, ought to be honored.

It follows that in contrast to Buchanan and Brock, whose stand allows for autonomy to be overridden in favor of the patient's basic interests, Beauchamp and Childress, despite viewing autonomy as merely prima facie, nonetheless give it priority even over well-founded concerns about the incompetent's best interests. What is more, in situations when a surrogate decision does not seem to threaten these interests, one still does not know, according to Beauchamp and Childress, when, if ever, the autonomy of a formerly competent patient ought to be balanced against some sort of objective criterion, nor does one see in what relation this former autonomy stands to the actual proxy's autonomy which plays the dominant role in both interpretation and implementation of the patient's prior wishes. Therefore, I hold that despite declaring the avoidance of the "ghostly autonomy" which hinders SJS,

Beauchamp and Childress do not seem to avoid it themselves, as their account of PAS seems to pose more problems than it actually solves.

Interestingly, Veatch does better with respect to the two-autonomy problem by using and explaining the notion of autonomy extended, whereas Buchanan and Brock provide grounds for a resolution of the conflict between the two autonomies by resorting to the category of "basic interest," which has its genesis in the law. Beauchamp's and Childress's approach, somewhat like that of Veatch, is also deficient because of its insensitivity to the way surrogate decisions are actually handled by practitioners in an everyday, real-life medical context. As stated above, the objective criterion which is overlooked in their understanding of PAS constitutes an indispensible reference point for physicians who deal in practice with proxy recommendations.

5.1.3 The Best Interests Standard (BIS)

BIS has been designed to address those who have never expressed their preferences. It is thus in contrast to the standards of substituted judgment and pure autonomy, which essentially concern individuals who when competent communicated values and wishes regarding a medical decision to be made after competency ceased. As is well known, the majority of the patients who fit this classification are the permanently incompetent. In its classic legal version, BIS "requires the surrogate to decide according to her views of what would benefit the patient most" (Berg et al. 2001, 114). Agreement has been reached among present-day bioethicists that the surrogates' views on the matter ought to reflect not their subjective preferences but rather, to invoke one representative explanation, the determination of "the highest net benefit among the available options, assigning different weights to interests the patient has in each option and discounting or subtracting inherent risks or costs" (Beauchamp and Childress 2001, 102).

The assumption underlying BIS is twofold. First, it is presumed that what is beneficial to the patient lies in his or her interest, which is interpreted to mean the greatest net benefit that it is possible to attain. Second, by benefit in general one understands what serves the patient's well-being. With its aim being to safeguard the individual's well-being, BIS aspires to an objective status, which is to say it implies that it is feasible for the surrogate to arrive at the knowledge of an objectively verifiable option that best serves the ward's well-being.

It is precisely at this point that the situation becomes quite complex. As Veatch points out, the ward's well-being may have at least two dimensions (Veatch 2003, 51–53). First, there is a medical well-being that amounts to what seems best from the purely medical point of view. Even if agreed upon, the most promising medical option still needs to be proved to represent a genuine benefit to the patient. Because health, although important, is virtually never the only goal or value that humans cherish and take into account in medical settings, not every form of medical well-being can be considered beneficial. For example, side effects or the burdens of a proposed treatment may negatively affect other aspects of human life valued by the patient to such a degree that they are not offset by any benefit arising from a particular medical procedure. Even supposing that this evaluation was never articulated or made known to the proxy, it does not follow that it does not exist.

The second type of well-being to be considered with respect to BIS is called by Veatch "total". In short, medical factors alone cannot be regarded as the entirety of the patient's well-being. Besides the medical elements, total well-being encompasses also social, psychological, economic, religious, aesthetic, and legal aspects, to name just a few. Typically, various persons allocate different priorities to each of these groups and may sacrifice one in favor of another. Competent patients or research subjects are considered free to decide how to pursue their total well-being and how much weight to give to its various constituents. But regarding the individuals to which BIS ordinarily applies, achieving this purpose presupposes an arduous effort and indeed, one that might easily turn into a Sisyphean labor.[92]

Moreover, even if it is assumed that under certain rather extreme circumstances such as permanent and profound incompetence the well-being of an incompetent can be confined to its medical aspect, it still remains enigmatic how to determine the medical benefit itself. According to Veatch, no fewer than four medical ends compete for the title of benefit: preventing death, curing diseases, alleviating pain, and promoting the well-being of the patient (Veatch 2003, 53). Not only may these four tasks be impossible to satisfy simultaneously but more importantly they may openly conflict with each other. Various practitioners and competent patients obviously come up with different ways of resolving these problems. Yet in my judgment no satisfactory guidance is to be found in today's professional literature on the appropriate basis for resolution in cases when BIS is to be utilized.

To complicate matters further, even if it seemed agreeable that medical well-being alone, rather than total well-being, sufficed for the implementation of BIS, and even if a proxy presented a credible idea of what in a given case should count as a medical benefit for the ward, one more quandary would remain. In order for the highest net benefit required for BIS to be determined, some sort of balancing between anticipated benefits and harms must be carried out on the part of the proxy.

No consensus exists, however, on what balancing tactic should be used to achieve this goal. At least three strategies present themselves as possible alternatives. The first, called arithmetic and favored for instance by Beauchamp and Childress (Beauchamp and Childress 2001, 102) as well as Buchanan and Brock (Buchanan and Brock 1989, 123), can be traced to the utilitarianism of Jeremy Bentham. In line with this strategy, "a decision maker should consider the amount of benefit to each person affected and then the amount of harm. The envisioned harm should then be subtracted from the benefit to reach a net figure for that individual ... Then the whole process should be repeated for every alternative course being considered—for example, every treatment option. The decision-maker ought to choose the option with the greatest net aggregate good consequences" (Veatch 2003, 55). The second method relies not on arithmetic but on geometric calculus, which usually comes down to determining the ratio between the benefits and harms associated with each medical option under consideration. Obviously, these two approaches are capable of yielding quite disparate results. But there is also a third strategy, predicated upon the assumption that no harm should ever be done to the patient. Usually referred to in its Latinized wording "primum non nocere" "it implies pursuing good for the patient only if one has ruled out the possibility of harming" (Veatch 2003, 57). Taken literally, this approach considerably decreases the number of morally justifiable medical options; it might even eliminate nearly all of them.

While presenting these three methods of harm-benefit balancing, Veatch advocates none of them. Yet, since he relates maximizing "the patient's net welfare" (Veatch 2003, 113) to the principle of individual (Hippocratic) utility, his approach seems most closely associated with that of Bentham and the arithmetic strategy outlined above. One real difference, however, is that "while Hippocratic utility focuses on the individual patient, classical utilitarians would consider the effects on all parties" (Veatch 2003, 55).

Still, even if not a classic utilitarian, Veatch remains utilitarian in his approach to BIS, as he describes the principle of Hippocratic utility in terms of increasing net benefits and decreasing harms. In its original subjective variant the principle relies solely on a practitioner who evaluates the outcomes (benefits and harms) of a proposed treatment. By contrast, the objective version of the principle takes into account the consensus of the medical community with respect to the indicated procedures based on "peer review, utilization review, quality assurance, outcomes research and treatment protocols" (Veatch 2003, 49). Despite calling this an objective version, however, Veatch admits that "there is, in principle, no way that quantifying benefits and harms and combining those two into an integrated estimate of the value of alternatives can be made objective" (Veatch 2003, 58). Therefore, what Veatch appears to mean by objective is simply less subjective, or objective only to a certain degree.

On the other hand, Veatch's admission of the necessarily subjective character of the process of determination of the highest net benefit for the incompetent, although somewhat indirect, sheds valuable light on the nature of BIS, which is alleged by contemporary mainstream bioethics to be objective. Put simply and in a strict sense, the word "objective" as commonly used is a misnomer in this context. To use it correctly the leading bioethicists would have to refer to a reliable standard of objectivity and then employ it as the basis for surrogate decision making. In this way, an approach to SJS already practiced by physicians, that is, presupposing comparison to an objective standard, would be expanded to allow for considering the total well-being of the patient, as opposed to only medical well-being. Also, the strategy proposed by Buchanan and Brock of balancing SJS against basic interests would acquire more ethical weight if there were an objective standard at hand.

5.2 Major Issues

In the *Encyclopedia of Bioethics* entry on surrogate decision making composed by Brock, as well as in *Deciding for Others,* authored by Brock along with Buchanan, two values, self-determination and the well-being of the patient, have been recognized as the foundation on which the practice of informed consent, including its proxy version, builds (Brock 2004, 2484; Buchanan and Brock 1989, 29–41). In light of the review of standards of surrogate decision making just conducted, it could technically be inferred that while SJS and PAS aim at protecting the former value, it is BIS that safeguards the latter. This conclusion is only partially true, however. These authors believe that self-determination and well-being cannot easily be divided, as the latter is deemed to be honored and sought by the autonomy-based substituted judgment of a proxy and not only by BIS. To put it another way, "self-determination is instrumentally valuable in promoting a person's well-being" because "what best serves the person's well-being often depends on the particular aims and values of that person" (Buchanan and Brock 1989, 37). Thus self-determination, frequently referred to in ethics as autonomy, emerges as the most fundamental, although not the only, value essential in surrogate decision making.

Beauchamp and Childress articulate the predilection for autonomy that is ubiquitous in modern mainstream bioethics when they elevate the principle of respect for autonomy to the rank of "the only foundation" (Beauchamp and Childress 2001, 103) for the pure autonomy standard which, as already noted, refines and thereby closely relates to substituted judgment. In this view, it is respect for the once-competent ward's autonomy that should govern the proxy's decisions. Nevertheless, the issue still remains open with regard to patients permanently lacking competence. Beauchamp and Childress actually refrain from providing a justification for BIS, which seems typical of today's professional literature.

As shown in the preceding subsection and contrary to common belief, determining the highest net benefit among obtainable options renders BIS essentially subjective. The notions of "benefit" and "best interest," which are critical to this standard, are generally presented as encompassing what is reasonably believed to serve the incompetent's good. Yet neither of the terms "good" and "reasonable," as they are usually employed, extends beyond what is purely subjective, which is to say that even if they are not arbitrary, they are still at best merely subjectively verifiable within a broader group of people. Any direct reference to the objective, in the proper sense of the word, is rarely found. But, as noted above, the objective simply cannot be reduced to a set of subjective opinions, even if they do not appear unreasonable. In point of fact, such subjective opinions, no matter how plausible, cannot be more than intuitive in character if divorced from a truly objective standard. By the same token, any attempt to glean what is good for the incompetent person within BIS will be based solely on intuition, or at best on some loosely conceived social agreement, unless a reliable theory of an objective good is offered.

It is precisely the latter, however, from which numerous mainstream bioethicists recoil, as in Buchanan's and Brock's approach. Not only does it not presuppose a theory of an objective good but in fact it offers no theory of good whatsoever. This is not to say that Buchanan and Brock entirely ignore such theories, though. In fact, as a result of surveying contemporary moral philosophy they single out what they call "three broad and promising theories of well-being or the good for persons." The first theory, known as hedonist, maintains that "the only thing that is good for a person is having conscious experiences of a specified, positive sort." According to the second theory "what is good for persons is for them to have their desires or preferences satisfied to the maximum extent possible over their lifetimes" (Buchanan and Brock 1989, 31). Third, there is an ideal approach, with its claim being that things prove "good or bad for a person independent of whether the person him- or herself will undergo positive conscious experience while having them and independent of whether the person prefers them, even after his or her preferences have been suitably corrected" (Buchanan and Brock 1989, 33). Despite allowing for some degree of objectivity, this third approach does not determine the content of ideals, leaving this aspect open to thinkers of various intellectual preferences. Thus the ideals, although they appear to have a degree of objectivity, are construed as initially content-free and therefore they are unavoidably prey to subjective interpretations.

No further explanations seem needed at this point to see what a limited and non-objective notion of good the three theories itemized by Buchanan and Brock feature. No matter which one of these theories they elect to act on, the proxy decision-makers will receive very little guidance. As a matter of fact, even if mistaken in the perception of the good of their wards, the surrogates may still find themselves justified in choosing that "good". It follows that in sharp contrast to an observation made in the second chapter of this study, the three theories of good seen by Buchanan and Brock as representative of today's moral philosophy assume that it is ultimately not the good that shapes people's preferences but rather it is their preferences that decide what is good. Even in applying ideal theory, it is preferences that are most likely to fill ideals with content. Aware of this, Buchanan and Brock overtly admit that well-being is "substantially subjective" for all three theories (Buchanan and Brock 1989, 35). Consequently, even if less subjective than SJS or PAS, BIS nonetheless never achieves objectivity, thus calling into serious question the extent to which the incompetent patient's good (one may use here other terms such as best interests or benefits) is really protected. If it is not, then the main lines along which mainstream accounts of surrogate decision making proceed must be reviewed and redefined. I believe that this is the case as it is impossible to safeguard something reliably while not knowing what it actually is.

The predominance of the subjective element in all three standards of surrogate decision making renders them content-thin, if not content-less, and hence not properly equipped for providing a solid and trustworthy guidance for the proxies. On occasion an attempt has been made to compensate for the absence of an objective notion of good and rationality within BIS by reference to quality-of-life judgments. Yet the concept of quality of life also is not free from subjectivity. In consequence, even the good of human life does not seem to be always desirable and worthy of protection.

The final, and highly significant, problem affecting the mainstream accounts of surrogate decision making is that they restrict their attention to the therapeutic setting. As a matter of fact, very little has been said thus far in professional literature on how SJS, PAS and BIS work when employed in proxy decisions for research subjects. Obviously, insofar as therapeutic research is examined, its correlation to the fully therapeutic milieu in most cases will be direct and straightforward. Still, one cannot be equally optimistic in the nontherapeutic research. While I intend to discuss the intricacies of nontherapeutic research at greater length in a subsequent chapter, I now turn to the critique of the notion of autonomy on which SJS, PAS, and in part BIS rest.

Chapter 6
Critique of Proxy Consent Standards

As the previous chapter has shown, the principle of autonomy has been endorsed as the primary rationale for proxy decision making as far as SJS and PAS are concerned. The purpose of this chapter is threefold. First, I will scrutinize critically the principle of autonomy as it is put forth and applied to proxy decision making by its major proponents. Second, I will refer to fundamental flaws in the way this principle is understood, with the main focus on its anthropological and ethical consequences. Third, I will show that autonomy, construed as pure self-determination, underlies not only SJS and PAS, as typically proposed, but also BIS, as a result of autonomy's decisive impact on the way the notion of good, undergirding the patient's or research subject's best interests, is perceived.

6.1 Status of the Principle of Autonomy

While the prominent position that autonomy occupies in present-day bioethics cannot be questioned, it is important to note that various scholars, including those whose views were presented in the preceding chapter, take different, albeit not necessarily uncorrelated, stands on autonomy's binding status. Thinkers such as Beauchamp and Childress opt for applying prima facie principles, including the principle of respect for autonomy, in the bioethical framework, whereas others either disapprove of them entirely or call for their refinement. As Pellegrino and Thomasma observed, "many remedies are offered to replace, prioritize, complement and supplement prima facie principles" (Pellegrino and Thomasma 1993, 189).[93]

One way of making the set of prima facie principles more workable in clinical and research settings is to ground them in an overarching principle. Another approach employs various ranks of priority among the principles, the practice usually referred to, after John Rawls, as lexical ordering. How does autonomy work through these two methods?

One influential example of the former approach, which raises autonomy to the position of an all-encompassing principle, is to be found in an early work of H. Tristram Engelhardt (see Engelhardt 1986).[94] The strong predominance of autonomy permeates his account and reveals itself particularly well when set against

G. Mazur, O.P., *Informed Consent, Proxy Consent, and Catholic Bioethics*, Philosophy and Medicine 112, DOI 10.1007/978-94-007-2196-8_6,

beneficence. Indeed, the tension between these two principles has marked bioethics virtually from its very beginning. It has also marked more or less conspicuously the previous chapters of this study, including the preceding one with its focus on standards for proxy decision making.

It is also, at least to some degree, along the lines of the tension between autonomy and beneficence that the second way of supplementing prima facie principles, the one employing lexical ordering, unfolds. The way Veatch understands medical ethics provides an example (Veatch 1981, 296–98). First, Veatch brooks no overarching principle because he finds it "most implausible" and likely to be of "not much help even if it could be found" (Veatch 1981, 298; Veatch 2003, 171). Second, Veatch distinguishes between consequence-maximizing principles (individual or social beneficence, individual or social nonmaleficence) and duty-based principles (individual: autonomy, fidelity, veracity, refraining from killing; and social: justice) that in his view make up essential constituents of respect for persons. He then proposes that the principles within each of these two groups (but not across the groups) are co-equal and as such subject to balancing. Moreover, the group of consequence-maximizing principles is construed as lexically ranked below the group of duty-based principles, from which it follows that, for instance, in case of a conflict between individual beneficence and autonomy, the former can never prevail, quite a different conclusion from that of Beauchamp and Childress (see Veatch 2003, 172–74; Veatch 1981, 298–305; see also Veatch 1984, 39).

But given all this, the question of how to balance the principles within each group still remains unsettled. Although he presents a number of insights on the matter, Veatch does not offer a clear, guiding solution. In cases when competing duty-based principles result in an impasse, the suggestion is that a lower-ranking principle of beneficence or nonmaleficence can be utilized as a "tie-breaker" (Veatch 2003, 173). Even with this avenue of escape, however, the workability and reliability of balancing are still open to examination, because balancing (1) relies on intuition and (2) allows for the possibility of a virtually infinite number of options that can potentially outweigh even the strongest moral beliefs (Veatch 2003, 172).[95] These issues notwithstanding, Veatch still believes in balancing within each of the two groups of principles he puts forward, with that balancing further refined by lexical ranking between two groups.

From the standpoint of this chapter's task the most relevant question regards the way in which balancing methods affect the principle of autonomy. Although in general Veatch treats autonomy as equal to other individual nonconsequentialist (duty-based) principles, these other principles may take precedence over autonomy. Veatch nonetheless contends that this will occur very rarely, if ever, in research (Veatch 1996a, 51). To an even greater degree this "rarity clause" applies to the principle of justice, which is a principle based on social duty and which may be in contention with autonomy. As Veatch phrases it, "while it may be true that in theory the principle of justice could constrain autonomy, it is critical to emphasize how unusual those circumstance would be" (Veatch 1996a, 54). Much the same can be admitted of social beneficence which, although capable of taking precedence over autonomy under certain circumstances such as "compulsory immunizations and quarantine for public health" (Veatch 2003, 174), could rarely if ever do so in research. In addition, the overriding power of autonomy is clear in regard to individual beneficence. As already mentioned, on the strength of lexical ordering as Veatch sees it, autonomy always trumps individual beneficence when the two principles are in conflict with one another.[96]

In contrast, Beauchamp and Childress find no justification for employing lexical ordering among their four principles. Consequently, they allow for the principle of respect for autonomy to be overridden by individual beneficence in some cases (see e.g. Childress 1990, 15). The same pertains to nonmaleficence and justice. Yet, as noted above, to posit equal status of all four principles, as Beauchamp and Childress do, invites reservations. In short, even if theoretically able to be overridden by other principles, autonomy more often than not wins out over these other principles in the application to concrete cases. As pointed out by Plich, almost all cases featured by *Principles* are settled in favor of autonomy (Plich 2006, 133). Thus, even if they officially resist utilizing lexical ordering and thus appear, at least initially, more cautious about the relevance of autonomy, Beauchamp and Childress surreptitiously introduce it, which in turn positions their approach quite close to that of Veatch.

Nevertheless, Veatch's support for autonomy in surrogate decision making still appears stronger. Although Beauchamp and Childress favor PAS over SJS in regard to once-competent and now-incompetent individuals, unlike Veatch they do not attempt to extend these individuals' factually non-existent autonomy, nor to provide for familial autonomy in the case of the never-competent whose family members are available to become proxies. In fact, Beauchamp and Childress never advocate autonomy as an immediate rationale for proxy decision making for permanently noncompetent individuals. These differences notwithstanding, neither Beauchamp and Childress nor Veatch absolutizes autonomy in the way Engelhardt does in his early writings, which is to say they do not establish autonomy as one supreme, all-encompassing principle on which virtually all ethical content ultimately rests. The same holds true for Buchanan and Brock, who refer quite noncommittally to self-determination, another word for autonomy (Buchanan and Brock 1989, 37). Hence it is not clear to what extent, if any, they would support a prima facie standing for autonomy. What is clear, however, is that they certainly provide no support for its absolute pre-eminence. In fact, they call for some kind of balancing when self-determination comes into conflict with well-being. As seen in the previous section of this chapter, they do not allow SJS an independent standing and postulate that it ought to be balanced by the basic interests limitation. So, even if it is not entirely clear to what extent Buchanan and Brock approve of a prima facie standing of the four principles included in Veatch's and in Beauchamp's and Childress's approaches, they certainly do not go as far as to endorse an absolute version of this principle.

Without doubt, the views on autonomy endorsed by Beauchamp, Childress, Veatch, Buchanan and Brock embody a greater degree of nuance and subtlety than the notion put forth by Engelhardt in his earlier work, which of course makes them harder to criticize. In one place, Childress refers to "several major criticisms" and complains that "many of those criticisms are displaced because they are (perhaps deliberately) not directed at the most defensible conceptions of the principle of respect for autonomy" (Childress 1990, 12). Unquestionably, the strong, overarching notion of autonomy initially endorsed by Engelhardt is the least defensible and even if it could be plausibly argued that both Veatch and Beauchamp and Childress do not actually stray as far from it as they assert they do, I nonetheless assume that there is a fundamental difference between his notion and theirs. Accordingly, in order not to ascribe too much to their views, and to avoid setting up an "autonomy straw man," I have decided not to include Engelhardt's approach to autonomy, which is absolute and indeed clearly self-defeating, into this critique. This is not to say, however, that this approach, although no longer shared by Engelhardt himself, does not find supporters in present-day bioethics, yet given its radical nature it is perceived as unacceptable by the majority of scholars.[97]

6.2 Autonomy as Pure Self-Determination

Regardless of the differences existing among them, the accounts favoring non-absolute accounts of autonomy share one tragic flaw: they refer to autonomy in a minimalist way, i.e., in such a way that nearly all content is discarded. In fact, the only content-bearing aspect welcomed in autonomy as thus construed is self-determination. This approach, as thoroughly demonstrated by Servais Pinckaers, came into existence in the thought of William Ockham and was later adopted, more or less consciously and consistently, by numerous modern thinkers. Freedom reduced to pure self-determination denies any intrinsic, intimate ties with the good and therefore Pinckaers calls it freedom of indifference (see e.g. Pinckaers 1995, 327–53). Such freedom, not influenced by anything apart from itself, stands in stark contrast to the freedom of excellence, which is not only not indifferent to the good but actually assumes it and is ordered toward it. Freedom of indifference involves autonomy as mere self-determination. It sees an act of choice as being ultimately dependent upon human will which, without reference to the good disclosed by reason, chooses arbitrarily between options. Freedom of excellence, in turn, recognizes the intrinsic bond between the will and the good that constitutes the will's object. It follows that the will does not create the good but rather presupposes it (Pinckaers 1995, 331).

Building upon Pinckaers' analysis of human freedom as well as on the contributions made by Iris Murdoch (Murdoch 1970) and D.C. Schindler (Schindler 2002), Plich depicts the Achilles' heel of reducing autonomy to merely self-determination as threefold.

6.2.1 Anthropological Consequences

First, this reduction has serious anthropological consequences, as it denigrates reason up to the point of radically renouncing it. Sound anthropology recognizes that acting morally presupposes both reflecting (knowing, deliberating) and choosing. The faculty responsible for the former is reason, whereas it is the will that exercises the latter. According to St. Thomas Aquinas, reason and will "unite to make the act of choice" (Pinckaers 1995, 331). In truth, willing good presupposes knowing it, as one cannot want what one does not know. Ralph McInerny puts the matter well when he says after Aquinas that the will is "a rational appetite, a desire that follows on cognitively grasping something good" (McInerny 1997, 69). In other words, "knowledge of the good specifies the act of will, and . . . in the order of specification, mind moves will by presenting its object to it" (McInerny 1997, 71). Consequently, will "cannot not will the good . . . Goodness itself . . . is the necessary object of the will" (McInerny 1997, 72).

It is critical to emphasize that good perceived in this way influences the will intrinsically, which is very different from any kind of external, invasive determination. Unfortunately, this distinction has been missed altogether in modern accounts that understand autonomy as pure self-determination divorced from good, that is, as if it could not will the good. Along with treating all good influencing the will as a potential threat restraining its operation, this relegation of freedom to its merely formal aspect divorces it from content and inexorably results in comprehending the human power to choose as relying solely on an indifferent will. Therefore, in an anthropology that adopts the notion of human will thus crippled, "reason and the entire human cognitive faculty, which recognizes what is good, and what are the best ways of its realization, becomes the enemy of the will and its self-determination, that is, its autonomy" (Plich 2006, 269).

Although they do not avoid incorporating the split between reason and will into their approach, Beauchamp and Childress do not seem to embrace fully its consequences. In fact, they still refer throughout their analyses to the process of reasoning based on an understanding of relevant information that is somehow capable of molding human choices and as such plays a critical role in the informed-consent process. It is not clear, however, how this process actually comes about and what concept of reason underlies it. However it is referred to by our authors, human reason does not function in their account "as a faculty presenting what is good to the will" (Plich 2006, 275).

The bifurcation between will and reason that is characteristic of modern ethics permeates the accounts of surrogate decision making previously examined in this study. Arguably the most striking definition ruling the faculty of reason out of the orbit of moral choice has been provided by Buchanan and Brock, who state that "a *deliberate choice* is an act of will or commitment" (Buchanan and Brock 1989, 116). Can deliberation be an act of will? Of course, it cannot. As adduced above, it is an act of reason, about which Buchanan and Brock remain silent. Although their understanding of choice contradicts itself, it still proves valuable because it reveals that the notion of the human will as indifferent to reason ends up taking over the role of reason. Attendant to this is the fact that the word "reason" does not meaningfully exist in any of the approaches to proxy decision making that were presented above. If it is referenced, this is done in a merely technical way that does not grant to reason the position of the faculty of the human person that introduces good to the will.

In Veatch's vocabulary, "reasonable" is used to signify something sufficiently congruent with choices judged reasonable by other people. No mention of objective good indicated by reason is ever made. In his critique of Veatch's concept of reasonableness, Ramsey notes that it relies heavily on social consensus, which explains why it carries so little objectivity. In the previous section, I have already critically referred to Veatch's definition of objectivity, which actually conditions it on subjectivity.[98] Ramsey enlarges upon this reservation and comes to the conclusion that what Veatch actually perceives by reasonable objectivity is in fact left to the patient's autonomy. Thus, "in the case of the competent adult patient the reasonableness standard turns out to be entirely subjective and arbitrary" (Ramsey 1978, 163). In consequence it is hard to see any reason why the refusal of important medical research therapy could not be based on whim or caprice. Interestingly, the human-rights documents scrutinized in the first chapter of this study, save *Nuremberg* and to a lesser degree *Belmont*, also fail to circumvent the trap of arbitrariness that, out of care for safeguarding autonomy, centers more on the self-expression of a patient or research subject than on his or her real good.

6.2.2 Autonomy and the Theory of the Good

The second issue presented in Plich's argument consists precisely in this: construing autonomy as nothing more than self-determination renders it abstract and arbitrary. In order for the will to retain and sustain its autonomy, it has to remain blind and random in its acts. Within such a perspective, human acts of choosing bear a close resemblance to tossing a coin, with voluntaristic and capricious planning and pursuit of goals being preferred to quality planning ordered to the real good of a concrete person.

Needless to say, this brings about numerous negative consequences not only in research involving human subjects but indeed in the entire domain of health care. To give but one example, the sound notion of counseling as a prerequisite for the informed-consent process, which was introduced earlier in this study, implies the existence of an objective good that ought to be discovered, rediscovered, or merely better comprehended by both physician/researcher and patient. Counseling thus conceived, however, becomes irretrievably lost when the will is understood as separated from its true object. Instead of being discovered or dwelt upon, the good is created. In the research milieu, it is by expressing their preferences that research subjects are expected to decide what is good for them. Accordingly, autonomous choices underlie and condition the content of the good. Even the medical good, which is the only objective aspect of the good to be found in the dominant present-day accounts of surrogate decision making, is typically considered good for the patient only insofar as it is chosen and approved. For this reason, Buchanan and Brock are not accurate in their articulation of one meaning of self-determination as "instrumentally valuable in promoting a person's well-being" (Buchanan and Brock 1989, 37).

In light of the analysis just conducted, there seems to be no question at this point that the influence of self-determination on well-being (that is on what is good for the person) reaches far beyond the merely instrumental. In fact, it proves critical in the accounts of proxy decision making prevailing in present day literature, as it undergirds the whole perception of good for the patients and research subjects.

This precedence that self-determination takes over the good is easily found in Beauchamp and Childress, who do not provide for any theory of objective good but at the same time, as Plich points out, cannot satisfactorily do without it. In effect, they tacitly incorporate the good under the more neutral-sounding term "interest." (Plich 2006, 267). Veatch's treatment of familial autonomy, by virtue of which surrogate decisions for never-competent persons with family are made, presents another example. As referred to in the previous section, Veatch defines what is good for such incompetents by recourse to the subjective notion of familial autonomy. In other words, in his account it is a family member, who autonomously defines and determines what is good for the incompetent (e.g., the incompetent's well-being or best interests). Interestingly, Veatch intuitively senses that autonomy construed in this manner cannot constitute a bottom-line foundation on which to build the ethics of surrogate decision making, but he nonetheless sidesteps this issue by claiming that regardless of its character it is certain that such a basis, more fundamental than autonomy, in fact exists.

It is critical to note that severing autonomy from the objective good, leading to the former taking precedence over the latter, ultimately undercuts the very possibility of freedom. Along the lines of the example offered by Fulvio di Blasi, "a man *does not choose* to be attracted to a woman (that is, to *know her as a good* or end), but because he is attracted he *can* then *choose* whether or not to bring her to the cinema, whether or not marry her, etc. And similarly for the other human goods." To put it another way, "if, prior to the choice, everything were absolutely indifferent or of equal value for man ... he could not choose" (Di Blasi 2006, 173). Consequently, knowledge of the good conditions the very possibility of freedom, not vice versa. This fact has been missed altogether by the proponents of self-determination which thus affects the accounts of proxy decision making previously presented. A tenuous and illusive concept of choice pervades those accounts, as it stands behind such terms as "preferences" or "wishes" to which they frequently refer.

The conceptual precedence of autonomy over good, coupled with defining the latter by way of the former, exerts a profound influence on BIS and its aspirations to an objective status. Yet, an evanescent notion of good derivative of autonomy cannot bear scrutiny and as such it undermines not only the ostensibly objective nature of BIS but also the advisability of the strategies balancing SJS/PAS against BIS. Much the same can be said of the medical approach to SJS depicted by Tonelli because the medical good constitutes neither the only aspect of the entire good of the patient or research subject, nor necessarily an overriding one. Without a sound concept of good rooted in human nature, a physician or researcher becomes merely a technician with his or her activity constrained to providing medical service at the patient's or subject's wishes, not even knowing to what degree, if any, this has a beneficial outcome apart from the purely medical one. Therefore, if Tonelli is right that for SJS to work satisfactorily it needs to be evaluated against an objective criterion, both BIS and its implementation must be predicated on a meaningful concept of good, which is missing in the approaches to proxy decision making presented in the previous section. Otherwise one is doomed to balancing the subjective against the subjective with no yardstick whatsoever by which to measure available options reliably and to determine the optimal one.

Needless to say, if its sole objective constituent is medical good, a system of ethics that upholds pure self-determination fails to provide reliable guidance for physicians and researchers. The point here is that every physician/researcher can arrive at the knowledge of what is medically indicated without resort to any ethical theory. If he or she admits of needing such a theory, it is usually because of the desire of acting morally well (that is, choosing the real good) in the midst of competing values and conflicting options that might merely appear to be good or that might be good only to a certain degree. If an ethics theory in general does not provide guidance on how to act morally well, or provides very little guidance in this respect, not only its usefulness but indeed its raison d'être stands in question.

To this charge Beauchamp and Childress respond as follows: "we regard disunity, conflict and moral ambiguity as pervasive features of the moral life. Untidiness, complexity, and conflict may be perplexing, but a theory of morality cannot be faulted for realistically incorporating these dimensions of morality" (Beauchamp and Childress 2001, 390). Certainly, at the level of incorporation of moral disorder, Beauchamp's and Childress' approach does very well. But at the same time, it must be emphasized that this approach, and indeed any other that does not considerably extend beyond simply recognizing "disunity, conflict and moral ambiguity" in human life, necessarily confines itself to a mere description of certain aspects of moral reality. Or, to be more accurate, it simply acknowledges what is commonly known anyway. A decent description might nevertheless prove to have a certain value but one needs to bear in mind that even the most thorough description does not easily result in understanding and guidance. More fundamentally, the way in which the facts are attended to, described and organized draws from prior philosophical presuppositions.

Beauchamp and Childress, however, remain largely silent on these presuppositions and even if they advert to some of them, this is accompanied by a consistent endeavor to limit their content to a minimum. Such a practice in turn affects the way in which Beauchamp and Childress describe moral life. It seems obvious that the lesser the content, the more limited the understanding and hence the more superficial the description. Thus, the manner in which these authors chose to do ethics results in my estimation not only in the lack of real guidance but also in a poor description of ethically important facts.

The dearth of reliable guidance coupled with overstating the descriptive dimension of ethical theory can be illustrated by the way the influences capable of overcoming autonomy to various degrees are presented. The question, of course, is how to glean which influence violates autonomy and which does not. As outlined earlier, Beauchamp and Childress handle this issue by putting forth three types of influence: persuasion, manipulation, and coercion. The last is defined by the reference to the threat of harm but, as pointed out by Plich, "the negative concept of harm is unthinkable without a positive account of the good" (Plich 2006, 281). Indirectly, this conclusion holds true also for manipulation and persuasion because their meaning as envisioned by Beauchamp and Childress turns on the way coercion is detailed. By way of recapitulation, they say that coercion results in the harm of controlling another, whereas persuasion never does. Manipulation in turn denotes an influence that is neither persuasive nor coercive. Hence, all three influences turn out to be traceable to the notion of harm, which, if devoid of intimate ties to good, bears no meaning whatsoever.

6.2.3 Autonomy and Intrinsic Goodness

The third charge Plich levels against reducing freedom to pure self-determination concerns an even more elementary area. In brief, this denigration undercuts the relevance and indispensability of intrinsic goodness, "that is, goods in themselves which become ends that are willed for their own sake" (Plich 2006, 270). One example of such a good is human life. In research, accepting risks that unavoidably lead to a grave and disproportionate harm to the subject's life or even to its loss can never be considered morally good and hence justified. But this conclusion remains unsupported when pure self-determination is granted a superior and isolated status by reason of which things become morally good to the extent that they constitute the object of autonomous choice, while prior to that choice no intrinsic good is deemed to exist. It is on this basis that options morally wrong per se may be legitimized.

Consequently, any competent patient or research subject can choose to do harm to himself and such choice, at least theoretically, ought to be respected. If any safeguards to offset this conclusion are offered, which indeed happens, for instance, on the institutional level, this is typically done with no reference to the notion of autonomy itself but rather because of a potential for abuse. Thus, as seen above, certain exceptions to the binding force of pure self-determination in surrogate decision making are considered justifiable but usually only on the grounds of providing patients or research subjects with protection against harm they might have not chosen if they were competent.

On occasion, the protection against harm comes down to questioning the competence of the patient/research subject if his or her wishes are judged unreasonable (in Veatch's understanding of the word). This issue has a critical bearing on proxy decision making, which, as noted previously, is typically justified by a lack of competence unless emergencies or other compelling factors warrant exceptions to informed consent. Hence competence functions as a gatekeeper for surrogacy and the condition sine qua non for resorting to proxies.[99] With this in mind, Beauchamp and Childress define competence via autonomy but they do not explain precisely the interrelation existing between the two terms. Even though they assert that "the criteria of the autonomous person and of the competent person are strikingly similar,"[100] one still does not know in what this similarity actually consists. Furthermore, since one does not know what competence actually is, it is impossible to construct a conceptual device by which to evaluate it. Indeed, Beauchamp and Childress come up with no such tool. Therefore, their all-too-brief assessment of competence does not provide significant help in the search for satisfactory safeguards for patients or research subjects when doubts arise regarding their competence.

6.3 Autonomy vs. Beneficence

At the outset of this study I critically referred to the fact that Faden and Beauchamp portray beneficence and autonomy as opposing each other. The issue was also invoked in the context of Engelhardt's early absolutist stand on autonomy. Now, in light of the analyses carried out in the current section, it becomes possible to demonstrate an ultimate rationale for the division in question. Basically, autonomy can be thought of as in opposition to beneficence for two reasons.

First, prima facie principles, regardless of who employs them, stand in potential conflict with each other only "because deprived of their essential relationship with the good they lose a common unifying point of reference, become free-standing and indifferent . . . , and they lack any intrinsic relations that mutually influence and systematically reconcile their contents" (Plich 2006, 279). Consequently, were they anchored in an objective notion of the good, the principles would be capable of working together in better understanding of the good as well as in its implementation in concrete cases.

Second, in many accounts including that of Faden, Beauchamp, and Childress, beneficence receives a paternalistic interpretation with no place reserved for the real good of the patient, which is jointly discovered by the patient/subject and his or her physician/researcher. In general, paternalism assumes that the physician knows best what is good, e.g., beneficial, for the patient, by virtue of which this good can be imposed on the patient regardless of his or her opinion. By contrast, beneficence implies doing good rather than imposing it. Therefore the real conflict takes place not between the patients' autonomy and their good or the doing of that good, but rather between patients' autonomy and some, certainly not all, physicians' paternalistic attitude. Pellegrino and Thomasma enlarge upon the error of confusing beneficence with paternalism and juxtaposing the former against autonomy as if it were the latter. This finally leads them to advance the principle of beneficence-in-trust, a remedy that reconciles and reunites autonomy with beneficence (see Pellegrino and Thomasma 1988, 54–55).

To comprehend more clearly the consequences of the pure self-determination model of autonomy, it is important to ask why Faden and Beauchamp tend to blur the line between beneficence and paternalism. In fact, the answer seems quite straightforward and has already been given. The major reason comes down to emptying beneficence of the content by reason of which it is set against other prima facie principles. Put differently, beneficence often becomes paternalistic because no good has been offered to undergird its meaning, so that typically it is others, most prominently physicians, researchers, or proxies, who attempt unilaterally to determine what is good for the patient or research subject.

Hence, the strife between autonomy and beneficence has its genesis in a potential dissonance between the subjective perception by the patient/subject of what is good and what is thought of as good by the doctor/investigator. This is why two autonomies are often in conflict and indeed prove irreconcilable in the approaches that endorse pure self-determination. The rule of thumb that a justifiable limit for one's autonomy can be set only by the autonomy of another, which is put forward by some as a remedy to this problem, obviously is not sufficient to bring about a solution. Indeed, there is no obvious resolution as long as free choice means arbitrary choice separated from the good. The result of this is that any option autonomously chosen remains valid. Unless one resorts to the concept of good, which invalidates at least some choices as incompatible with the good and thus morally wrong, no remedy against the arbitrariness of two conflicting autonomies is to be found.

Chapter 7
The Golden Rule and Proxy Decision Making

It is clear from what has been said in the previous section that using autonomy conceived as pure self-determination as a rationale for surrogate decision making reveals a serious deficiency. In the current section, I intend to show that the decision-making rationale can be identified with the Golden Rule, as formulated in the Christian Bible and then referred to by St. Thomas Aquinas. The Rule provides a sound justification for acting as a proxy, as it presupposes a meaningful account of the good and of virtue and also recognizes basic inclinations of human nature, along with the need for friendship and love.

7.1 In Search of a Rationale

It is noteworthy that while a broad agreement exists among scholars, including Catholic bioethicists, on the justifiability of proxy consent in therapeutic research, there is no such agreement on its rationale. The Magisterial documents examined in this study are not an exception in this regard. The *Charter for Health Care Workers* provides substantial clues on how to deal with the matter, especially by emphasizing the importance of personal choice that presupposes a profound concept of freedom irreducible to sheer, self-determinative autonomy. Nevertheless, it touches explicitly upon neither proxy consent nor its conceptual basis. Even though the *Charter* refers to presumed consent and allows for its employment in therapeutic research, the document does not qualify to what degree, if any, this type of consent applies to surrogate decision making. Likewise, the *Charter* does not advert to proxies themselves or to possible ways of their appointment, yet it stipulates that patient's relatives "should be informed about ordinary interventions and involved in the decision making when there is question of extraordinary and optional interventions" (n. 74).

The Ethical and Religious Directives for Catholic Health Care Services differ from the *Charter* in that they provide for advanced proxy directives with the assumption being that the proxy's decisions are "faithful to Catholic moral principles and to the person's intentions and values, or if the person's intentions are unknown, to the person's best interests" (n. 25). Faithfulness to Catholic moral principles calls for a

G. Mazur, O.P., *Informed Consent, Proxy Consent, and Catholic Bioethics*, Philosophy and Medicine 112, DOI 10.1007/978-94-007-2196-8_7,

recognition of the patient's or research subject's good, not limited to his or her preferences, which in turn requires that best interests are viewed and determined on the basis of the good, not vice versa. In addition, as already noted, the *Directives* propound conjoining formation of conscience with counseling aimed at the true good of a patient or research subject, an essential factor in genuine surrogate decision making. These advances notwithstanding, no plain rationale for proxy consent in therapy or research has been proposed by the document.

In a similar vein, Catholic philosophers and theologians dealing with medical ethics have to date not broached the subject in question. No basis for proxy consent in therapeutic research is to be found in such authors as Edwin F. Healy (see Healy 1956) and Gerald Kelly (Kelly 1958). Even if more articulate on the matter, Orville N. Griese (Griese 1987) and David Kelly (Kelly 2004) do not engage in a quest for such a basis. By the same token, although he asserts that in therapeutic research, "it is within the competence of the parent or a duly constituted guardian, in accord with their responsibility for the welfare of the incompetent patient, to give informed consent" Thomas J. O'Donnell (O'Donnell 1996, 122) leaves the underpinning of this responsibility unarticulated. Kevin O'Rourke in his *Primer for Health Care Ethics* (O'Rourke 2000) and both O'Rourke and Benedict Ashley in their fifth edition of their *Health Care Ethics* (Ashley, DeBlois, and O'Rourke 2006), are also among the roster of Catholic bioethicists who do not pronounce on the matter. Nevertheless, the fourth edition of *Health Care Ethics* features a notable exception in this respect. Ashley and O'Rourke assert that "guardians have responsibility for wards because of the principle of human dignity in community which affirms that no person can achieve fulfillment without sharing in the common good and contributing to it" (Ashley and O'Rourke 1997, 351). The principle of human dignity in community "rests on a basic human need that must be satisfied to attain true happiness, namely, the need of every person for society" (Ashley and O'Rourke 1997, 215).

In their account of basic human needs, Ashley and O'Rourke draw from Aquinas' recognition of the four goods characteristic of human nature, that is, those goods to which all human beings find themselves naturally inclined: (1) "preserving human life," (2) procreation and the rearing of offspring, (3) knowledge of the truth about God, and (4) living in society (ST, I-II, q. 94, a. 2).[101] Although predicated upon the latter good, the principle of human dignity in community also encompasses other basic needs as it "sums up the true goal of human life: integral human fulfillment in relation to God and neighbor" (Ashley and O'Rourke 1997, 215).

Yet it remains unclear in this account how exactly the principle works with respect to proxy decision making. In other words, it is insufficient to say that since both a proxy and his or her ward are humans inclined to the same set of goods, with the emphasis on the good of living in a society, a proxy therefore ought to decide on behalf of a research subject. Likewise, it is not enough to admit that the principle demands that "all ethical decisions, including those in health care, must aim at human dignity; that is, the maximal, integrated, satisfaction of the innate and cultural needs, biological, psychological, ethical, and spiritual, of all human persons." While at a general level there is not much to question in this formulation, one needs further guidance at the more particular level of surrogate decision making. Interestingly, Aquinas himself was quite precise at this point and aware of the fact that the list of four basic goods derivative of human nature and enumerated above is not exhaustive, in the sense that it needs to be translated into more specific principles. Thus, for instance, the good of living in a society implies honesty and precludes deception, disrespect for one's property, and injury, to cite but a few examples.

The principle guiding humans in their mutual relationships and social endeavors to which St. Thomas explicitly refers is the Golden Rule (Eberl 2006, 14). It is important to note that as used by Thomas, the negative formulation of the Rule ("Do not to another what thou wouldst not were done to thyself"; cf. ST, Supp. q. 65, a. 1, ad. 8), echoing the book of Tobit, chapter 4, verse 15 ("What you hate, do not do to any one"),[102] finds its affirmative parallel in the logion of Jesus in the Sermon on the Mount: "whatever you wish that men would do to you, do so to them" (Mt 7:12). Grisez notes that although "it is sometimes argued that the affirmative formulation substantially extends the negative one," the accuracy of this claim hangs upon "the precise interpretation one gives the various formulae" (Grisez 1997a, 213).

William E. May, the only American Catholic bioethicist who, to my knowledge, directly addresses the problem of a rationale for proxy decision making in therapeutic research and takes the Golden Rule to be precisely this rationale, uses the formulation that encompasses both negative and positive dimension of the Rule: "We are to do unto others as we would have them do unto us and not do unto them what we would not have them do to us" (May 2008, 220).[103] Although May never elaborates on it at a greater length, the reason for his choice of Golden Rule becomes more evident in light of the analyses thereof offered by Grisez and utilized below.

7.2 Golden Rule, Reason and Virtue

Closely related by Grisez to the fifth mode of responsibility[104] and hailed as its "most famous formulation" (Grisez 1997b, 866), the Golden Rule expresses the principle of impartiality (Grisez 1997a, 189). In therapeutic research partiality can be generated, to give but one example, by a conflict of interests. Because of this, Nicholas Tonti-Filippini reminds us that research interests should always be honestly revealed so that the proxy can evaluate them in order to "ensure that the integrity of the decision to treat is not compromised by research or commercial interests" (Tonti-Filippini 2007, 757). For Grisez, being impartial implies the virtue of fairness without which, in my estimation, the Golden Rule remains merely an unworkable ideal.

Of course, the problem here does not confine itself to the virtue of fairness only. In fact virtue as such, which is to say, a virtuous, good moral character, comes into focus. Therefore, even if one verbally conforms to the concept of the Rule, it does not follow that one is capable of conforming one's actions to it as well. Indisputably, it is virtue, or moral character, to put the matter more broadly, that makes it possible for human agents to act in accordance with the Rule.

In this regard, even if it has not been expressed by Grisez himself, it is worth noting a link between virtue and reason as identified by St. Thomas Aquinas. To put it briefly, Thomas explains that one can approach virtues in two ways; "first, under the aspect of virtuous; secondly, as such and such acts considered in their proper species." Then he goes on to say that it is according to the former way that "all acts of virtue are prescribed by the natural law." In other words, the "natural inclination to act according to reason," which all human beings share, means "to act according to virtue" (ST, I-II, q. 94, a. 3). Thus, being virtuous proves reasonable in the sense of an objective quality rooted in human nature.

This conclusion, obviously, stands in sharp contrast to the meaning that reasonableness receives in secular accounts of proxy decision making. The reasonableness characteristic of those accounts, impoverished as it is in terms of content, cannot reach beyond what is subjectively found agreeable within a group of people. Consequently, a reasonable-person standard, typically referred to as a basis for objective judgments, easily slips into both legal and ethical fiction. If the Golden Rule is to be more reliable than what people happen to judge or feel is right, it needs to build on virtue embedded in what is truly good for man. Accordingly, in Aquinas' view, virtues develop the natural inclinations and bring them to perfection (Pinckaers 1995, 336). Thus construed, virtues not only enable one to apply the Golden Rule successfully, but also provide the Rule with content. Indeed, it does not suffice for a proxy to have a moral character strong enough to apply to the ward what he thinks he would have wanted the ward to apply to the proxy if the roles were reversed. What the proxy needs as well is knowledge of an objective good for both, that is, for the proxy and also, by recourse to the Golden Rule, for the ward, as indicated by human nature.

In research settings, as pointed out by Thomasma and Pellegrino and already referenced earlier in this study, the virtue of particular significance is prudence, "the chief practical habit of man's intellect." Prudence manifests itself in "reasoning well about, and commanding, good moral actions" (Bourke 2008, 184–85). Certainly, the Golden Rule cannot do well without it, nor can any standard of reasonableness. Therefore, I propose that the reasonable-person standard that occurs frequently in the accounts predicated on purely self-referential autonomy be replaced by a prudent-person standard which returns the objective character of reason to its proper role in the structure of the human act.

The link between reason and virtue identified by St. Thomas cuts against the grain of emotions and feelings identified in Grisez's account as a major source of partiality and therefore as a hindrance to the Golden Rule.[105] Thus, to say that a virtuous proxy acts reasonably means, among other things, that he or she possesses a habitual capability of avoiding an emotionally based prejudice or bias. This in turn is momentous for surrogate decision making in where feelings on the part of proxies, especially when they happen to be close friends or family members of an incompetent, may easily eclipse reason. The accounts of proxy decision making examined earlier do not safeguard against emotional bias although they refer to it critically on occasion.

Elaborating further on partiality with relation to the Golden Rule, Grisez declares that the meaning of the former does not boil down to selfishness or egoism. In fact, "possibly more common, and certainly as *unreasonable* [as selfishness and egoism], is *allowing* one's choices to be shaped by personal likes and dislikes, jealous love of one's family, group prejudices, culturally established patterns of bias, and so on" (Grisez 1997a, 212; emphasis added). It is noteworthy that Grisez not only calls shaping one's choices according to personal likes and dislikes unreasonable, he also applies the same judgment to allowing for that which bears more directly on acting as proxy. None of the approaches to proxy decision making based on self-determination satisfactorily attends to this issue. On the contrary, as previously pointed out, personal likes and dislikes are often permitted and honored as decisive in the application of SJS or PAS. In fact, this in itself seems unavoidable as long as reasonableness remains caught in the trap of subjectivity that readily confines itself to mere emotions and feelings.

If the Golden Rule is to remain truly resistant to partiality and thus provide reliable guidance for proxies, it must integrate emotions and feelings and it must be understood objectively. This is to say that it is to be understood in a way that presupposes acting in accord with human nature, i.e., in harmony with its basic inclinations (the four goods plus the inclination to act in accord with reason, both invoked above). Further, this understanding must incorporate the characteristics of human nature expressed by the precepts of natural law as well as developed by virtues. The Golden Rule thus construed overcomes two common objections noted by Alan Donagan.

According to the first objection Donagan refers to, the Golden Rule "excludes the possibility that it may be right to do anything to another which you would hate to have done to you." According to the second, "the Golden Rule prima facie fails to condemn any action which affects the agent alone (as suicide may) or any action between consenting persons, to which there is no other party" (Donagan 1977, 58). In my opinion, both charges can be met by referring to the objective good. Regarding the first charge, one could hate having a leg amputated as a means of dealing with gangrene but this would not change the fact that the amputation was precisely the good to be done in such a situation, with the possible exception of some circumstances making the whole procedure unduly burdensome. Likewise, using the notion of objective good settles the latter objection by disqualifying suicide or other actions that are wrong per se (intrinsically evil) from among the spectrum of morally good options.

The notion of good undergirding a sound comprehension of the Golden Rule also bears upon balancing benefits and harms, an issue critical in research, including its therapeutic type. According to O'Donnell's concise formulation reflective of the Catholic position, therapeutic research cannot be legitimate unless two conditions are met. First, "there is no proven remedy available," and second, "the experimental procedure holds out a notably greater hope of benefit than any proven remedy and the danger associated with it is at least moderate and controlled" (O'Donnell 1996, 122). The latter condition pertains to assessing benefits and harms, the practice necessary, as Grisez emphasizes, for applying the Golden Rule. In the accounts of proxy decision making scrutinized above, this practice, even if directly invoked and desired, ultimately fails because of the lack of a yardstick with which to measure, compare, and weigh the benefits and harms. Indeed, to be able to offer such yardstick one needs to allow for an objective good translatable into an objective hierarchy of goods that serve as axiological points of reference indispensible for a meaningful employment of any balancing tactics.

The Golden Rule has an important role to play in both discovering and applying the objective good as it relates the good first to the one who makes a judgment for and on behalf of someone else. Thus understood, the Golden Rule does not conform to any approach that constricts its purpose to merely putting oneself in the other's place without securing objective values, embedded in human nature. With such an approach the Rule would become essentially prima facie, which is to say its employment does not guarantee a morally good outcome. In consequence, its power to guide is lost and the objections, including those articulated by Donagan, become much harder to refute.

7.3 The Golden Rule, Friendship, and Christian Revelation

To begin with, it is worth reiterating that the two formulations of the Golden Rule noted above are to be found in the Bible. In his work, Grisez shows that divine revelation not only gives rise to the Golden Rule but also provides for its most profound fulfillment. As already discussed, the impartiality to which the Golden Rule gives expression implies the virtue of fairness. Fairness, which can also be called disinterestedness (i.e., acting on one's own interests in a fair, uncompromised way), constitutes an important facet of justice (Grisez 1997a, 212). Both justice and fairness belong to the attributes of God who "practices no partiality and is utterly and absolutely faithful; he is unconcerned with justice to himself, for his own goodness is absolute." Every Christian is called to embrace by grace this selfless disinterestedness whose fullness resides in God and has been revealed to mankind in His Son, Jesus Christ. Building upon these fundamental premises, Grisez proposes that the fifth mode of Christian response, which is "faith's specification and charity's fulfillment of the fifth mode of responsibility," means "to be merciful according to the universal and perfect measure of mercy which God has revealed in Jesus" (Grisez 1997a, 644). This formulation corresponds to the fifth Beatitude: "Blessed are the merciful, for they shall obtain mercy" (Mt 5:7).[106]

Referring to the New Testament parable of the merciless servant (see Mt 18:23–35), Grisez states that "Christians, who are aware of their need for God's mercy and have received it, are bound by the Golden Rule itself to extend this same mercy to others" (Grisez 1997a, 645). Another parable, namely that of the Good Samaritan (Lk 10:30–37), can be invoked at this point, as well, because it fits precisely the theme of mercy in the context of surrogate decision making. The Good Samaritan acts out of compassion and shows mercy to the wounded and robbed man. His role combines that of a proxy (he makes a decision on how to care for the man and then, after providing first care, gives directives to an innkeeper) with that of a physician (he himself bandages up the man's wounds). In both cases mercy underlies the Good Samaritan's actions.

Interestingly, the Parable of the Good Samaritan is placed immediately after the commandment of love of God and neighbor has been recalled. In fact, Jesus conjoins the command: "You shall love your neighbor as yourself" (Mt 22:39) with the Golden Rule (see Mt 7:12). The former command, as Pope Benedict XVI avers in his encyclical letter *Deus Caritas Est,*[107] cannot be read in separation from the one that immediately precedes it, i.e., the one that Jesus calls "the great and the first" (n. 18) and which says: "You shall love the Lord your God with all your heart, and with all your soul, and with all your mind" (Mt 22:37). Therefore, the Golden Rule ought not to be detached from the command of love of God. The close connection between the Golden Rule and the double love commandment has been lucidly shown in the Gospels. In Matthew, Jesus says of the Golden Rule that it summarizes "the law and the prophets" (Mt 7:12) and he reiterates this in regard to the commandments of love on which "depend all the law and the prophets" (Mt 22:40) (Grisez 1993, 307).

Living out love in human relationships presupposes friendship. The natural inclination to live in a society on which the Golden Rule rests finds its foundation, in line with Pinckaers' explanation, in "human need for friendship, affection and love." This need, common to all human beings, has been recognized and described long before Christianity, for example by Aristotle. In the *Nicomachean Ethics* Aristotle regards friendship "one of the most indispensable requirements of life" which either is virtue or "involves virtue" (n. 8.i; after Aristotle 1962). Furthermore, he distinguishes three kinds of friendship: one based on utility, a second on pleasure and a third on willing good for another. About the latter Aristotle has this to say: "it is those who wish the good of their friends for their friends' sake who are friends in the fullest sense, since they love each other for themselves and not accidentally" (n. 8.iii). Thus, the latter type of friendship turns out to be essentially identical with love manifesting itself in the disinterested wishing of good for the friend's sake. Echoing Aristotle, Pinckaers concludes that "friendship is the special type of love that befits moral excellence, the love of friendship, which consists in loving another person or loving the good in itself and for its own sake" (Pinckaers 1995, 437). It is precisely this kind of good for an incompetent person that is sought in proxy decision making.

The natural inclination to the social life can be identified not only by the human need for friendship but also, as Aristotle points out in the *Politics*, by a characteristic of all humans, the power of speech (n. I.I.10; after Aristotle 1944). This power distinguishes man from other non-human living creatures and fundamentally conditions the possibility for communication on which any human friendship builds (Pinckaers 1995, 433).

An example of incorporating the nexus between friendship and power of speech into proxy decision making has been offered by MacIntyre. Acting as proxy, in his view, can be done only on the grounds of past friendship. Yet, in order to speak for those who cannot speak for themselves and "only have a voice ... if someone else speaks for them" one needs to learn first how to speak for oneself. The encumbrances constraining this kind of learning are, for instance, an infantile and unconscious need for approval which makes one unreasonably conform to the opinions of others, or the opposite tendency based on the contrary need to be seen as never conforming, which leads one to constant disagreement with the opinions of others. Such hindrances, nonetheless, are possible to overcome and, as MacIntyre points out, "it is by having our reasoning put to the question by others, by being called to account for ourselves and our actions by others, that we learn how to scrutinize ourselves as they scrutinize us and how to understand ourselves as they understand us" (MacIntyre 1999, 148). Learning intelligibility with regard to others usually goes hand in hand with acquiring the ability of being intelligible to ourselves. Most importantly, however, when we "make ourselves accountable to others and treat ourselves as accountable to them, we have to become able to assume the other's point of view, so that the concerns to which we respond in giving our account are the ones that are in fact genuinely theirs" (MacIntyre 1999, 150). In other words, as we work on becoming accountable for others by speaking for ourselves, we also learn how to speak for others. Plainly, these two abilities prove critical in both forming friendships and in acting as proxies. In fact, a mere capacity to discern between them can hardly be overestimated in the face of a frequent temptation to project the proxy's own wishes and values on the ward. Moreover, accountability and intelligibility, indispensible when one is acting as proxy, presuppose a good moral character and virtue. This means that one needs to work on one's character with the aim of attaining moral integrity. A morally disintegrated proxy who never learned how to account for herself or himself intelligibly and consistently cannot reliably account for another.

Attaining moral integrity requires an objective good. MacIntyre harbors no illusions in this regard and points out that even if the questions one receives from others do not explicitly bring up the word "good," they always imply it and can be traced back to it. Therefore, adequate responses to these questions call for the good, by which one's answers become not only intelligible to others but also justifiable. Accordingly, MacIntyre's approach, although it can be applied directly to SJS because of its reliance on past friendship, can also be utilized in BIS, thanks to its openness to the good. Much the same can be said of the Golden Rule, which serves as a means by which morally intelligible and accountable proxies apply the good, objectively understood, in concrete, real-life situations.

Although MacIntyre does not bring it up explicitly, both accountability and intelligibility can be raised at a higher, in fact the highest attainable, level by recourse to the twofold commandment of love. Characteristic of all human beings, the natural need for friendship finds its fulfillment precisely in love of God and neighbor. As Pinckaers summarizes it, for both Thomas Aquinas and the Fathers of the Church, "the relationships nature has established between us and other creatures are an extension and a participation in the supernatural ties that unite us to God at the source of our nature and being" (Pinckaers 1995, 435). It is precisely by means of these ties that humans can build friendships and also love each other at a supernatural level.

Pope Benedict puts the matter well when he asserts in *Deus Caritas Est* that by virtue of "an intimate encounter with God" one learns to look on another not simply with human eyes and feelings, "but from the perspective of Jesus Christ," who invited all mankind to friendship. Jesus loved and died for all. This universal perspective helps us understand that "in God and with God" one can love those whom one does not like or even know (n. 18). In the realm of proxy decision making the latter conclusion holds extreme importance, especially with regard to BIS. As far as SJS is concerned, past friendship, developed either at a solely natural level, in the case of a nonbeliever, or on the grounds of a personal relationship with God, comes to the fore. In both BIS and SJS love constitutes the most profound justification for all kinds of decision making.

Consequently and fundamentally, the Golden Rule can be viewed as a rationale for proxy decision making in therapeutic research only insofar as it presupposes and expresses love. It is in true love that the Rule finds its ultimate, objective meaning. No better safeguard to protect research subjects can ever be offered. No more efficient yardstick by which to balance benefits and harms is to be employed. Finally, no more accurate recognition of human good pertinent to both proxies and research subjects can be achieved than that embedded in love.[108]

Chapter 8
Preliminary Considerations on Proxy Consent in the Nontherapeutic Situation

In this chapter I will first consider different types of nontherapeutic research integrated into a broader framework of basic research taxonomy. Following this, four categories of research subjects particularly prone to abuse in nontherapeutic setting will be surveyed. These categories include: psychiatric and medically handicapped patients, prisoners, children and the unborn. Under the fourth category I will discuss the issues of embryonic stem cell research, genetic therapy as well as prenatal and pre-implantation screening. These considerations lay the groundwork for the next chapter, in which the debate on the justifiability of proxy consent in nontherapeutic research is presented.

8.1 Nontherapeutic Research and Basic Research Taxonomy

The goal of this section is a concise demonstration of where in research the therapeutic-nontherapeutic distinction applies, with the main focus on which types of research are precisely nontherapeutic. As a matter of fact, numerous research projects include both therapeutic and nontherapeutic elements, so that at times it is extremely difficult, if not impossible, to classify the entire project simply as therapeutic or nontherapeutic. In similar fashion, and at an even more rudimentary level, it might be necessary to differentiate within a project between those of its constituents that are components of standard medical practice and those that are experimental by nature and hence cannot be classified as standard interventions. In fact, research studies frequently feature a mixture of both (Weijer 1998, 854). Since different ethical measures need to be taken with respect to standard and innovative practices, the distinction in question proves critical. Furthermore, if the innovative practice happens to be part of a research endeavor, a further question must be asked regarding its therapeutic and nontherapeutic components. As has already been demonstrated earlier in this study, if a research proposal contains nontherapeutic elements, more stringent ethical rules ought to be applied, regardless of the number of coexisting therapeutic constituents.

G. Mazur, O.P., *Informed Consent, Proxy Consent, and Catholic Bioethics*, Philosophy and Medicine 112, DOI 10.1007/978-94-007-2196-8_8,

8.1.1 Basic vs. Clinical Research

To begin with, a fundamental distinction between basic research and clinical research must be recognized. The former "is usually laboratory based and includes studies at the cellular level, and of immunity and pathogenesis." In contrast, the latter, "is often conducted with patients in a medical setting, such as a hospital, and is designed to obtain better information on the natural history or pathogenesis of a condition that may lead to improved strategies for diagnosis, treatment or prevention of a disease" (Serrao 2003, 191). Clinical research involves human beings as subjects and can be defined as "a structured process" that "proceeds in a systematic way to examine clinical conditions and outcomes, to establish relationships among clinical phenomena, to generate evidence for decision making, and to provide the impetus for improving methods of practice" (Portney and Watkins 2000, 4). As indicated above, clinical research may contain both therapeutic and nontherapeutic constituents depending on its objectives and design.

8.1.2 Descriptive Research: Case Studies and Case Series

This type of research grows out of the recognition of the fact that "a description of interesting, new, and unique cases is necessary to build a foundation for clinical science and as a means of sharing special information among professional colleagues" (Portney and Watkins 2000, 267). A classic example of the research called *case study* focuses on the description of individual conditions. When the focus expands to "group, institution or other social unit, such as a particular school, community, or family" (Portney and Watkins 2000, 267), the research in question becomes a *case series*. A case series "is essentially a record of clinical practice in which the clinician's experience with one or more similar patients is reported" (Weijer 1998, 855). As such, it can be used as an evaluative tool with regard to standard medical practice and innovative therapy, and even in research, including nontherapeutic research, although perhaps not to the same degree as in therapeutic settings. Because of its limited nature, the case series is considered by some not to be a separate research design but rather a tool applicable in numerous research endeavors. Furthermore, the records of case series typically handle rather small numbers of subjects that may lead to "errors of interpretation" on account of "chance events" (Schaffner 2004, 2330). This is also the reason that the case series are known as susceptible to bias.[109] Hence, they "may mislead when the reporting of cases is incomplete ... or more subtly, if clinicians—intentionally or unintentionally—select preferentially certain types of patients for inclusion in the study, for example, patients with the best prognosis" (Weijer 1998, 855).

8.1.3 Exploratory Research: Cohort and Case-Control Studies

This type of research implies "the exploration of relationships to determine how clinical phenomena interact" (Portney and Watkins 2000, 124) and can be illustrated by reference to observational analytical studies which in turn fall under a broader category of *epidemiological* research. The latter aims at the enhancement of the health of populations and may feature both longitudinal[110] and cross-sectional designs.[111] Observational analytical designs are employed when "enough is known about a condition to allow testing of hypotheses about the association of specific risk factors and outcomes" (Portney and Watkins 2000, 324). This in turn often takes the form of "searching for the causes and effects of observed phenomena" (Bompiani 2003, 101).

Observational analytical research can be divided into two types: *cohort* and *case-control* studies. In the former, subjects are recruited on the grounds of their exposure to a risk factor in focus. For instance, in a study of the impact of smoking on the incidence of lung cancer, smoking counts as a risk factor, whereas the lung cancer is an outcome (see Weich 1998, 284).[112] Cohort studies are prospective,[113] from which it follows that "exposure status must be ascertained before outcome is known" (Weich 1998, 284). By contrast, in case-control studies "subjects are chosen on the basis of having an outcome of interest or not, that is, they are either a 'case' or a 'control'" (Weijer 1998, 855).[114] Accordingly, these studies are retrospective[115] because the exposure factor is taken into consideration only after the determination of the outcome. Obviously, the therapeutic-nontherapeutic distinction does not apply in a meaningful way to observational research. This distinction, as referred to in professional literature, implies direct or indirect intervention, medical in character, as opposed to merely the observation which is characteristic of both cohort and case-control studies.

8.1.4 Experimental Research: Clinical Trials

Clinical trials, especially in their randomized variant, epitomize experimental research. Described as "a collection of study designs specifically conceived for the rigorous evaluation of the safety and efficacy of new medical treatments" (Weijer 1998, 855), clinical trials differ from observational studies in that they necessarily imply the use of "one or more intervention techniques" (Friedman, Furberg, and DeMets 1998, 2). These techniques may include testing prophylactic, diagnostic, and therapeutic compounds (Friedman, Furberg, and DeMets 1998, 2).[116] In addition, clinical trials have two critical characteristics: (1) efficiency in avoiding biases and (2) application of methods of statistical analysis in the field of empirical research, with the former frequently amplified by the latter. By definition, a clinical trial is "a prospective study comparing the effect and value of intervention(s) against a control in human beings" (Friedman, Furberg, and DeMets 1998, 2). In other words, researchers "compare the outcomes for patients getting one treatment with those who do not. This allows investigators to separate the treatment's effects from other influences" (Kopelman 2004b, 2335). It follows that at least two groups of subjects must be employed in a clinical trial, one of them being the control group.[117]

The basic understanding of controlled clinical trials (CCTs) has already been offered in the first chapter of this study in the context of placebo-based studies carried out in the underdeveloped countries with the aim of enhancing the prevention of maternal-fetal HIV transmission. There I referred to two types of control groups: placebo and standard-treatment. Control provided by these two groups is concurrent. Hence, their full names are respectively placebo concurrent control group and standard (or active) treatment control group. In addition, a no-treatment concurrent control group and a dose-comparison control group may be selected and employed in CCTs. A historical control group can also be used when comparison is made between the present and past patients' outcomes.[118] Regardless of the number of groups that comprise a particular CCT, it must be emphasized, after Loretta Kopelman, that "the goal of clinical trial methodology is to compose and treat groups as similarly as possible except for the one feature under study. Investigators attempt to identify other features that are likely to affect outcomes and to stratify or distribute individuals with those features equally between groups" (Kopelman 2004b, 2334).

Often, in order to moderate as far as possible the subject-selection bias and to secure the highest ascertainable comparability between the groups, randomization techniques are used.[119] Put briefly, "in the simplest case, randomization is a process by which each participant has the same chance of being assigned to either intervention or control" (Friedman, Furberg, and DeMets 1998, 61). One advantage of random subject allocation lies in the fact that it precludes personal preferences of research subjects and investigators (Bompiani 2003, 108). More specifically and with regard to research subjects, randomization considerably reduces the impact of personal habits, interests, and other factors, including physical characteristics and psychological traits, on the research outcome. In other words, randomization in CCTs strives "to combat bias in group assignments by distributing individual characteristics whose effects are unknown fully among the study aims to minimize their influence" (Kopelman 2004b, 2335). Randomization thus understood overcomes bias only to a degree, however. This is because, even if randomly selected, the subjects know to which group they belong, thus introducing an element of bias. Consequently, randomized Clinical Trials (RCTs), far more resistant to bias than their non-randomized counterpart, become even more impervious to bias when so-called blinding is applied in their execution.[120]

While randomization targets primarily the bias related to subject selection, blinding procedures aim at the elimination of "measurement bias" (McKibbon, Eady, and Marks 1999, 35). To put the matter simply, randomization minimizes the influence of the subjects' characteristics, which are unrelated to their outcome-expectations (health, habits, etc.), whereas blinding addresses precisely the issue of expectations by which the appraisal of the outcome is likely to be distorted. In single-blinded designs, "subjects do not know their group assignment, thereby minimizing the effects of their beliefs and expectations about the different modes of treatment" (Kopelman 2004b, 2335). When double-blinding is used, "neither the participants nor the investigators responsible for following the participants know the identity of the intervention assignment. Such designs are usually restricted to trials of drug efficacy" (Friedman, Furberg, and DeMets 1998, 85).[121]

Double-blind designs are commonly regarded as the most reliable or even as ideal (Friedman, Furberg, and DeMets 1998, 82). Nevertheless, in some circumstances double-blinding can be extended so that not only the subject and investigator but also "the committee monitoring response variables is not told the identity of the groups" (Friedman, Furberg, and DeMets 1998, 86; see also McKibbon, Eady, and Marks 1999, 36). This extension makes for triple-blinding.[122] In triple-blind trials "the health professional, the patient *and* the study personnel, including the data management staff, do not know which treatment is active therapy and which is placebo or standard therapy until final data analysis is complete" (McKibbon, Eady, and Marks 1999, 36).

As already mentioned, CCTs typically involve putting to the test novel treatments and that in turn entails examining potentially therapeutic, yet unproven, agents, such as drugs or vaccines, a prerequisite for approval leading to market release. Here are the four major phases of which CCTs consist.

Phase I, typically conducted on a rather small group of healthy volunteers, aims at determining pharmacokinetics (absorption, distribution, metabolism and elimination) as well as pharmacodynamics of a new drug, with the focus on "dose-responses, and in the case of vaccines, immune responses." The task involves attaining the "confirmation ... of the verified responses in animals" (Bompiani 2003, 104) as well as gleaning "the maximum dose that can be tolerated by participants" (Serrao 2003, 194).[123] Determining the relationship between dose and safety lies at the heart of Phase I and it may be referred to as "dose-escalation, dose-ranging, or dose finding" (Piantadosi 2005, 224). Obviously, human responses indicative of dose and safety, although predictable on the basis of prior animal studies, have their own specifics and may differ from one individual to another. Furthermore, although Phase I studies may use similar methodologies for animal and human subjects, with human subjects the researcher primarily seeks "the appearance or not of subjective symptoms (nausea, headache, etc.) that are not verifiable using laboratory animals" (Bompiani 2003, 104). It is on account of its employment of healthy humans as subjects that Phase I is nontherapeutic in nature.

Relying on "the information regarding the safe dosage range obtained from the phase I studies," in Phase II, "the compound will be administered to patients suffering from the target disease." In brief, what is sought in Phase II comes down to "evidence of the efficacy of the medicine against the specific disease" (Serrao 2003, 194). In more concrete terms, the objective is "to evaluate any biologic activity or effect and to estimate the rate of adverse events" (Friedman, Furberg, and DeMets 1998, 4). Phase II trials may employ a fairly large number of patients. As such, they are in principle therapeutic because the subjects involved are patients suffering from the disease against which a new agent is being tested.

While Phase II studies establish both some level of efficacy of a tested compound and the lack of significant side effects, the next step, Phase III, investigates the efficacy of the agent as juxtaposed with standard therapies. Conducted in multiple research centers, Phase III studies may involve thousands of subjects. Their "experimental design is almost always a double blind, randomized in confrontation with a placebo or with a drug with an ascertained efficacy" (Bompiani 2003, 105). A Phase III study may be both therapeutic and nontherapeutic, depending on the group to which a research subject has been assigned. While being therapeutic for the subjects receiving either a new or standard drug, it is nontherapeutic for those enrolled in a placebo group.

After the drug has been released into market, the study continues, although typically at the merely observational level, in Phase IV. Controls, randomization, and blinding are not utilized at this stage. Monitoring of a new drug is necessary because certain side effects can be detected only when the agent is received by vast numbers of patients (Serrao 2003, 195) and for a relatively long period of time. Much the same can be said of "interactions with other therapies, or unusual complications" (Piantadosi 2005, 150). Furthermore, it is widely agreed that a better understanding of the way in which the agent works can be achieved outside the research center, in real-life conditions. Phase IV makes it possible, for instance, to observe unpredicted and unintended therapeutic properties of a new agent (Bompiani 2003, 105). These advantages aside, Phase IV trials may be conducive to gaining deeper knowledge of the effects the drug has on particular groups of patients, such as children or the elderly (Serrao 2003, 195). Because of its primarily observational nature, Phase IV escapes the therapeutic-nontherapeutic classification. Thus it may entail epidemiological designs, referred to in previous paragraphs.

8.2 Vulnerable Research Subjects

As is well known, research on human subjects regularly involves individuals who by reason of their mental, physical, or developmental condition are more vulnerable to exploitation and abuse than others. Recognition of this fact demands an even greater amount of caution to be exerted by all parties involved in the research enterprise, with the aim of ensuring sufficient protection of vulnerable subjects. This is especially an issue in nontherapeutic research, when proxy consent is secured, as may be the case of minors, or when consent is given in coercive environments such as penitentiaries and the research involves, for instance, participation in Phase-I clinical trials of new drugs with the incarcerated serving as subjects. In this brief presentation I will confine my attention to four groups of subjects particularly vulnerable in research: psychiatric and mentally handicapped patients, prisoners, children, and the unborn.

8.2.1 Psychiatric and Mentally Handicapped Patients

In their *Encyclopedia of Bioethics* article on research involving mentally disabled and mentally ill persons, Carl Elliott and Susan Parry identify and discuss two major issues characteristic of such research: evaluation of competence (decision-making capacity) and risk factors attendant to study designs (Elliott and Parry 2004, 1825). Both relate directly to informed consent, including its proxy type.

As far as evaluation of competence is concerned, in the interest of clarity one must note at the outset the difference, not immediately obvious, between mentally handicapped/disabled persons and those who are decisionally impaired. Although frequently used as if they were equivalent, the two terms nonetheless carry different meanings. In brief, the former does not necessarily entail the latter, which is to say that not all mentally disabled persons are decisionally impaired.[124] Decisional impairment renders an individual incompetent and thus unable to offer a valid, that is, free and accurately informed consent. Generally, "the more severe and chronic the mental disorder, the more likely that the person suffers from impairment of comprehension or decisional capacity" (Shamoo and Khin-Maung-Gyi 2002, 77).

Still, decisional impairment differs from cognitive impairment. To put the matter simply, while advanced cognitive impairment always precludes valid informed consent and hence calls for surrogacy, questions arise as to whether or not this also holds for its early and mild stages.[125] In addition, both decisional and cognitive impairment occur as either permanent or intermittent. Consequently, the degree of competence characteristic of a given subject may change over time and as such should be carefully assessed not only upon enrollment in research but also periodically during its course (Shamoo and Khin-Maung-Gyi 2002, 77).[126] Whenever compromised competence or its complete lack calls into question the subject's potential involvement in research, an additional inquiry on the research goals is in order. For instance, under certain circumstances it may be justifiable to enroll an incompetent patient in therapeutic research of an antipsychotic drug when the drug works so as to restore competence.

The issues of competence and informed consent become particularly difficult when mentally handicapped patients are institutionalized. These patients are said to be "especially attractive as research subjects because their medication, diet, and compliance with a study can be monitored and controlled easily" (Elliott and Parry 2004, 1827). Yet it is precisely because of their institutionalization that the patients in question may be abuse-prone to an even higher degree. As referred to early in this study, history has witnessed occurrences in which such patients were grossly exploited. The infection of mentally underdeveloped children with hepatitis virus at Willowbrook State School on Staten Island, New York, and the introduction of live cancer cells into the systems of unsuspecting chronically ill patients in the Jewish Chronic Disease Hospital in Brooklyn provide but two, albeit notably infamous, illustrations.

Evaluating the competence of mentally handicapped persons has at least two additional sets of hurdles. First, no agreement exists as to what degree of competence is sufficient for recognizing the validity of the person's consent to become the research subject. Should this degree be different in therapeutic and nontherapeutic research? Does an initial competence assessment require the same amount of decision-making capacity as subsequent reevaluations made in the course of research? No unequivocal answers to these questions can be found in contemporary professional literature. What is often emphasized, however, as for example in the

CIOMS Guidelines, is the need for seeking the assent and cooperation of incompetent persons involved in research, "to the extent that their mental state permits, and any objection on their part to taking part in any study that has no components designed to benefit them directly should always be respected" (commentary on Guideline 15). Yet, it is likewise accepted that an absence of assent or even overt dissent of an incompetent individual to research procedures may go unheeded when "the research is therapeutic, involves minimal risk, has been consented to by an appropriate surrogate, and is clearly in the best interests of the patient" (Elliott and Parry 2004, 1827). The CIOMS Guidelines specify that, for example, the lack of a "reasonable medical alternative" may legitimize overriding of the incompetent's dissent but only if the local law allows for that (commentary on Guideline 15).

A second issue is that "most proposed standards for assessing competence focus on the process of reasoning involved in making a decision rather than on the outcome of the decision" (Elliott and Parry 2004, 1826). Although not illegitimate, this practice bears considerable risks. For instance, by relating competence primarily to reasoning capabilities, one may easily factor out depression as impairing competence.[127] Even if it is chiefly a disorder of mood, seemingly unrelated to deliberation, comparison, and evaluation, depression still threatens competence, in particular in relation to risk-benefit assessment. Put simply, "if a person is depressed, he or she may be *aware* that a protocol carries risks, but simply not *care* about those risks" (Elliott 2003, 237). Moreover, the accounts of competence which concentrate principally on "rationality and reasonable deliberation may not be very helpful when the person making the choice has an affective disorder." Persons who suffer from depressive delusions, to give but one example, "may consent to hazardous research because they think they deserve to be punished" (Elliott and Parry 2004, 1827).

When the lack of competence is evident, the CIOMS Guidelines stipulate, in a manner representative of current professional literature, that before beginning research involving mentally ill and incompetent persons the investigators must ensure that (1) no such research may equally well be conducted on competent subjects; (2) the research aims at gaining knowledge pertinent to the medical condition of the subjects involved; (3) assent of each subject has been sought to the maximum possible extent based the subject's current capabilities; (4) a valid consent has been secured from a proxy (see Guideline 15). Notably, proxy consent figures as last on this list and becomes relevant only after the first three requirements are satisfied. As an additional safeguard, the National Bioethics Advisory Commission (NBAC) in the U.S. recommended that "all IRBs that regularly consider proposals involving persons with mental disorders should include at least two members who are familiar with the nature of these disorders and with the concerns of the population being studied." If such proposals are discussed only on occasion, IRBs ought to consult two independent experts (National Bioethics Advisory Commission 2003, 230).

When it comes to risk factors, a second major issue characteristic of research with handicapped subjects, it has been admitted by most contemporary authors that such research can be ethically permissible only insofar as it carries risk which is minimal or which only slightly exceeds the minimal level, the position reflected in various regulations, including the CIOMS Guidelines examined earlier in this study. This recognition notwithstanding, a far lesser degree of concurrence exists on how to determine minimal risk and what counts as a slight increase (Elliott and Parry 2004, 1826). Likewise, it is disputable whether or not the psychiatric research procedures are in themselves more risky than their medical counterparts.[128] Risk-related issues in research involving mentally handicapped subjects appear to reach the highest level of complexity in the three most controversial research designs: CCTs employing placebos, symptom-provocation studies, and relapse studies.

As far as the first design is concerned, providing a placebo to patients in a trial who suffer from more serious mental conditions such as schizophrenia or severe depression carries the potential of significant harm, including a substantial risk of suicide (Elliott and Parry 2004, 1828).[129] In symptom-provocation studies, also known as challenge studies, "a psychopharmaceutical agent or psychological challenge is administered to research subjects under controlled conditions to measure or observe the results" (DuVal 2004, 57). For instance, subjects diagnosed with schizophrenia may be administered a range of psychoactive agents to aggravate the symptoms of psychosis. In contrast to clinical trials, challenge research places subjects at risk without any prospect of therapeutic outcome. What is more, these studies differ from Phase-I CCTs in that subjects are recruited not from among healthy volunteers but from among ill patients. As such, the studies are not only nontherapeutic but they are also intended to produce deleterious effects on the condition of those already suffering from mental infirmities.

In relapse or washout studies, which are somewhat related to challenge studies, "mentally ill subjects are taken off their regular medications to determine whether they will relapse into their illnesses, how long it will take them to relapse or whether their health can be maintained without medication" (Elliott and Parry 2004, 1829). Washout research has proven typically to involve a great amount of risk, and as such it usually cannot bear ethical scrutiny.

8.2.2 Prisoners

Prisoners as a class of potential experimental subjects epitomize so-called captive populations that are suitable and easily accessible for research, but at the same time ordinarily not capable of making sufficiently free choices precisely because of their institutional confinement. It follows that securing valid informed consent in general along with retaining freedom from coercion in particular are the most demanding goals to attain when enrolling captive subjects in research. Overall, captive populations can be divided into those physically confined (prisoners, the institutionalized) and those confined "by virtue of their positions in power relationships with authority figures" (Emanuel et al. 2003, 227) (students[130] or military personnel[131]). Even a preliminary presentation of all these groups cannot be included within this subsection due to space constraints, and therefore the focus will be only on prisoners, the population with arguably the longest history of abuse and with a highly representative range of problems pertinent to the topic of involvement of captive populations in research.

Prisoners have been abused in nontherapeutic and largely involuntary experiments virtually since ancient times. In the U.S. it was only in the 1980s that the recognition of the unethical character of such research surfaced.[132] On the other hand, therapeutic research still enjoys the status of legitimacy and its nature is perceived broadly. In short, not only are the procedures that directly benefit the incarcerated subjects (therapeutic sensu stricto) often classified as therapeutic, but also those that serve the whole imprisoned population and may not directly benefit the subjects actually involved in a given project. In contrast, for other groups of research subjects, even if an indirect medical benefit is likely, such research remains on most accounts nontherapeutic. The National Commission for the Protection of Human Subjects of Biomedical and Behavioral Research played an important part in excluding nontherapeutic research from the array of justifiable options with incarcerated subjects. The Commission's efforts contributed to change in the social mindset as well as in public policies regarding the research in question. As noted in the first chapter of this study, the National Commission also authored the Belmont Report, and two of the *Belmont* principles have been recognized as particularly relevant to the issue of prison research.

The first principle is respect for persons, with its application being informed consent. The issue comes down to whether or not incarcerated subjects are capable of giving informed consent which is sufficiently free, that is, consent which is uncompromised by the coercive environment of the prison. Those who hold that this is possible emphasize that the majority of prisoners are competent individuals, attested to by the courts of law that in pronouncing sentence found them sufficiently competent to be held accountable for the crimes committed. Furthermore, the prisoners perform in prison various jobs for which they are frequently compensated. Even though the prison jobs usually carry some risk, sometimes even greater than that characteristic of medical research, nonetheless the prisoners are allowed to undertake them. Similarly, they should be allowed to serve as subjects in biomedical research if they choose to.[133] Others, however, contradict this claim by drawing attention to the fact that "the relationship of persons to their bodies is very different from their productive goods." In fact, "the former comprises their relationship to themselves." For that reason, there is a great difference "between activities in which impinging on a person's body is accidental or unavoidable, as in a job, and those in which it is the very purpose of the activity, as in experimentation" (Branson 2004, 2105). In consequence, because consenting to a job varies substantially from consenting to research, it is likely that while retaining enough freedom for giving consent to the former, the prisoner, nonetheless, is incapable of providing meaningful consent to the latter. In accord with this line of argumentation, certain authors disaffirm in principle the validity of informed consent under prison conditions. Perhaps in an ideal correctional milieu, it is sometimes argued, a valid, coercion-free informed consent to research could be reasonably expected of inmates, but it is hard to think of this expectation as legitimate in real-life prison settings, at least as presently known, based on the characteristics of the U.S. penal system (Branson 2004, 2105).[134]

The second *Belmont* principle often brought up in the context of prisoner involvement in research as subjects is that of justice. According to one view, reparatory justice demands using prisoners as subjects in research. Since the crimes they committed caused damage to society, this argument goes, the prisoners have a moral obligation to make reparation, which can come about through participation in research. Others, however, oppose this reasoning by pointing to the fact that society has already punished the perpetrators of crime by incarcerating them and there is no convincing reason to extend this punishment so that it encompasses participation in research. To help settle this and similar disputed matters the National Commission employed the notion of comparative justice according to which "like cases or classes are to be treated alike, and different cases or classes are to be treated differently" (Branson 2004, 2106).[135]

One illustration of how comparative justice may be applied is the issue of recompense for prisoners serving in research. Absolute recourse to justice would call for paying prison subjects the same amount for participation in research as for non-imprisoned volunteers. Such payments, however, would constitute a much more substantial amount than any recompense the inmates receive for performing other work in prison. Consequently, these payments would be overly attractive or even irresistible, and thus coercive. Accordingly, the National Commission proposed that while investigators ought to provide the same compensation for enrolling prisoners in research as they do with respect to free volunteers, the individual prisoners will be paid less, that is, about the same amount that they would receive otherwise for jobs done in prison. The remaining surplus can be used, for example, as contribution to "a fund for the general benefit of prisoners" (Branson 2004, 2106). This solution harmonizes with the essence of comparative justice because the prison population differs markedly from its free counterpart and as such can be treated differently.

Comparative justice also proves relevant to research involving prisoners because of its intimate link to subject selection, by which it in turn relates to distributive justice. As we have seen, there is widespread acceptance of the idea that burdens and benefits tied to participation in research should be allocated fairly among various classes of subjects. This means that no group should be overburdened with difficulties and risks on the one hand, and none can be underemployed or even excluded from participating in research and hence from accessing its outcomes on the other. With regard to that, no reason justifies excessive employment of prisoners as research subjects in, for example, Phase-I drug trials, with a simultaneously lesser involvement of other groups in society. By the same token, in nontherapeutic research the presumption that prisoners ought to serve as subjects cannot be supported. The comparative justice criterion helps to clarify that the opposite actually happens to be the case. Since prisoners are "in so many respects different from, and unequal to, the rest of society," they are treated justly "if they do not participate in research that does not benefit them directly" (Branson 2004, 2107).

Currently in the U.S. all federally sponsored research involving human subjects must comply with the Common Rule that comprises subpart A of 45 CFR 46 (Shamoo and Khin-Maung-Gyi 2002, 96; Friedman Ross 2006, 63). Three extra sets of protections have been affixed to subpart A, one of them (subpart C)[136] addressing prisoners participating in research as subjects. In accordance with this provision, studies carried out with prisoners must (1) include no greater than minimal risk in relation to "the possible causes, effects, and processes of incarceration, and of criminal behavior;" (2) constitute the "study of prisons as institutional structures or of prisoners as incarcerated persons;" (3) be relevant to the specific conditions of prison population; (4) have therapeutic design and purpose. Whenever the incarcerated subjects may be assigned to control groups, a consultation is required "with appropriate experts, including experts in penology, medicine, and ethics," and in addition a note needs to be published "in the federal register, of the intent to approve such research" (45 CFR 46.306).

8.2.3 Children

That children serving in research as subjects need additional protections sounds like a truism. But it is precisely here that the following dilemma forcefully presents itself: "with too few protections, children selected as subjects may be exploited. If the regulations impose too many protections, however, it may become so difficult to conduct research that the knowledge base for making good decisions for children will erode" (Kopelman 2004a, 391). The author of these words, Kopelman, refers to three approaches attempting to handle this dilemma.

The first approach, called libertarian, supports the enrollment of children in the same types of research in which adults participate. Children, it is argued, can be exposed to more than minimal risk even in nontherapeutic research if their proxies consent. Numerous scholars rightly counter this claim, however, and point to the fact that children do not constitute their parents' possessions, or more precisely, they are not things to be used in research but human persons whose dignity must be assured and put firmly above the interests of science or third parties. Therefore, making children undergo risky experimental procedures with no direct benefit for them can never be justified.

The second approach, called the Nuremberg solution, places itself on the opposite end of the spectrum and prohibits literally all research with children, even if it is therapeutic. Yet, as the CIOMS Guidelines assert, "the participation of children is indispensable for research into diseases of childhood and conditions to which children are particularly susceptible (cf. vaccine trials)" (commentary on Guideline 14). Still, the so-called "presumption against child research" requires that children not be employed as subjects in research that could just as well be done with adults (see Montgomery 2001, 173–74).[137]

A third, intermediate position, known as the "risk-benefit" solution, proposes that children can be involved in research with the proviso that it is therapeutic or "does not place them at unwarranted risk of harm, discomfort or inconvenience" (Kopelman 2004a, 392). This formulation opens the door for nontherapeutic research, but certainly in a more responsible way than the first solution does. As is the case with mentally handicapped patients, the permissibility of nontherapeutic research with children is typically viewed through the lens of the risk-benefit ratio it carries. In the U.S., subpart D of 45 CFR 46 allows for more than minimal risk in research with minors when it is (1) therapeutic (presents a direct benefit to the child) or (2) nontherapeutic but at the same time aimed at gaining "generalizable knowledge about the subject's disorder or condition." In the latter, more than minimal risk is restricted to "minor increase over minimal risk" (§46.404–407).

In both therapeutic and nontherapeutic research with minors, parents' permission along with the child's assent, if the child has reached the age of reason, must be obtained. With children involved in research, the term proxy permission seems more appropriate than proxy consent, because it can be argued that "consent signifies agreement to participate for oneself, whereas permission refers to agreement to participate for a third-party" (Friedman Ross 2006, 87). Informed consent for research involving children encompasses both parental/guardian permission and the subject's assent. As it is a particular instance of parental authority over a child, decisional power regarding both treatment and research can be justified on various grounds. Most fundamentally, it is assumed that parents know their children best and are determined to act for their good. This authority is not unlimited, however. When suspicions arise with regard to abuse, neglect, or exploitation of a child, parental custody may be revoked temporarily or permanently. As Kopelman writes, "when parental omissions pose an imminent danger to children, doctors, nurses, hospital administrators, and social workers have a moral and legal duty to seek a court order for proper care" (Kopelman 2004a, 388).

Informed consent for research with minors requires obtaining, whenever feasible, the subject's assent. Not surprisingly, the need for involving the children subjected to research in the decision-making process has been supported precisely by research conducted with children. For example, "social-science research has found that many children understand a great deal about their diseases and even their imminent death." What is more, "they sense when people are not truthful, and this can cause them to suffer by feeling isolated from discussion, decisions, and support" (Kopelman 2004a, 389).[138] At a more rudimentary level, seeking assent of a child stems from respect for his or her human dignity. As such it involves respect for who a child is and who he or she is becoming (see Friedman Ross 2006, 95).

CFR defines the child's assent as "affirmative agreement to participate in research" and adds that "mere failure to object should not, absent affirmative agreement, be construed as assent" (§46.402). The need for assent and the significance of that assent increase in direct proportion to the child's age and the complexity of the research project. A higher level of complexity furthermore implies a certain degree of decision-making capacity which includes the ability to reason about the nature and consequences of participation in the research (Leikin 2003, 258). Although far from sufficient for meaningful consent, this is nonetheless required in order for the child's assent to make sense. At the same time it is important to note that, while generally presumed incompetent to consent, some children, especially adolescents, may present adequate capacity for offering valid informed consent. Accordingly, the CIOMS Guidelines suggest that minors "over the age of 12 or 13 years are usually capable of understanding what is necessary to give adequately informed consent."

Even so, the child's "consent (assent) should normally be complemented by the permission of a parent or guardian" (Guideline 14).

A significant ethical difference exists between the enrollment in research studies of children who assent to participation, and the enrollment of unwilling minors, perhaps even despite their overt opposition to serving as subjects (Montgomery 2001, 177). Because of the dignity of children as human subjects, both their assent and their dissent call for recognition. Acting otherwise degrades children to being merely their parents' possessions and often results in undermining their trust in those otherwise participating in the research project (including parents or guardians). Yet the minor's assent and dissent still need to be distinguished from one another. Honoring the latter, for instance, can be looked upon as one of the purposes of seeking the former (Leikin 2003, 256–57). The CIOMS Guidelines utilize the term "deliberate objection" and emphasize that even children incapable of giving an assent may nonetheless be able to refuse participation in research. It is widely agreed in today's mainstream professional literature that such dissent must always have a binding force, even if parents or guardians consent to the research, unless "the child needs treatment that is not available outside the context of research" (commentary on Guideline 14).

8.2.4 Unborn Humans: Embryos and Fetuses

Research that involves human subjects in their prenatal phase of development, which commences at fertilization and ends at birth, is one of the most contentious issues being discussed in present-day bioethics. To begin with, it is important to invoke a number of biological facts. The prenatal phase breaks down into two stages: embryonic and fetal (Mauron 2004, 707). In scientific terms, an embryo means the product of conception until "eight weeks of gestational age when the name changes to fetus" (Tauer 2004, 713). The name change comes in the wake of the biological process of organogenesis which "involves the differentiation of the three germ-layers into specific tissues and primordial organs." By the end of the embryonic stage "many important organ systems are in place, at least in rudimentary form. The fetal phase is characterized by further differentiation and maturation of tissues and organs, as well as considerable growth" (Mauron 2004, 708).

These biological facts notwithstanding, it has become customary to use the term "embryo" as referring to the very beginnings of human development before the implantation takes place. Accordingly, the Code of Federal Regulations, in subpart B immediately following the Common Rule, defines fetus as "the product of conception from implantation until delivery" (45 CFR 46.202) and allows for research on fetuses when (1) it is therapeutic or (2) even though nontherapeutic, it entails no greater than minimal risk and "the purpose of the research is the development of important biomedical knowledge which cannot be obtained by any other means" (45 CFR 46.204). Subpart B thus recognizes the fetus as a human subject and attempts to provide a suitable framework for its protection. Yet, no safeguards whatsoever are offered with respect to the early, i.e., the pre-implantation embryo, which opens door for its unlimited utilization in research. In ethics the moral status of the early embryo has been heatedly debated up to the present time with some authors clearly denying, entirely or partially, its humanity or personhood.

The *Encyclopedia of Bioethics* lists four major lines of argument against the humanity of the pre-implantation embryo. According to the first line of objection, the possibilities of twinning and recombination (several embryos may conjoin so as to form one embryo) indicate the lack of individuation of the early embryo. The second argument reinforces this by emphasizing that until they enter the blastocyst phase, which typically occurs around five days after fertilization, human embryos are merely aggregations of undifferentiated cells rather than organized and individualized beings. Third, it is noted that the appearance of the primitive streak, i.e., "the groove along the midline of the embryonic disk" that "marks the beginning of the differentiation of the cells into the various tissues and organs of the human body" (Tauer 2004, 717) takes place only about 14 days after fertilization. Lastly, since in natural human procreation a large percentage of embryos are discarded, some take this as indicative of their non-human standing (Tauer 2004, 717–18).

Yet, upon closer inspection, all four of the charges arrayed against the humanity of the early embryo fail. Certainly it is true that if the early embryo lacks individuation and constitutes merely a collection of cells, it cannot be a human being, "for human beings are individual organisms of the species Homo sapiens. And if it is neither a human being nor an individual, then it is certainly not a person" (George and Tollefsen 2008, 149). However, as Robert George and Christopher Tollefsen demonstrate, reliable scientific evidence indicates that the early embryo actually is sufficiently unified to be regarded an individual (George and Tollefsen 2008, 151–58). Moreover and in regard to the first charge summarized above, no plausible data supports to a satisfactory degree the claim that either recombination or twinning precludes individuation. The early embryo, whether it fuses with another or continues development alone, "will proceed—because of its teleological orientation, and apart from accident or deliberate human intervention—to be a born human person and not any other creature" (Ashley and Moraczewski 2001, 190).[139] By the same token, twinning affects neither the early embryo's individual status as a human organism nor its teleology. Accordingly, George and Tollefsen emphasize that "every human embryo is a human being ... even if that embryo may give rise to twins" (George and Tollefsen 2008, 56). In further recognition thereof, Benedict Ashley and Albert Moraczewski draw attention to the fact that the very nature of twinning comes down to cloning, which implies that "organism A that was developing normally up to the point of accidental twinning continues its development alongside its somewhat younger clone, its monozygotic twin (organism B)" (Ashley and Moraczewski 2001, 198). Continual development of organism A confirms its individuality by reason of which it can also be distinguished from organism B. Moreover, "if it were true"—Ashley and Moraczewski go on to ask—"that as long as twinning remains possible the embryo must be simply a mass of independent totipotent cells (blastomeres), *then why are only one, two (or a very few) persons produced instead of as many as there are embryonic cells*?" (Ashley and Moraczewski 2001, 195, emphasis in text; see also George and Tollefsen 2008,

149–50). To my knowledge, no cogent explanation has ever been offered to this question.

The third set of arguments against the humanity of the early embryo revolves around the fact that twinning can occur in general only before the time that the primitive streak appears. For those who argue that as long as twinning is possible the human embryo counts as nothing more than a cluster of cells with no individuality,[140] the formation of the primitive streak defines the boundary of legitimacy for embryo-destructive research.[141] After the primitive streak appears such research is often said to be illicit (George and Tollefsen 2008, 46). Yet, as shown in the previous paragraph, in light of the scientific data, the occurrence of twinning does not contradict individuality. In addition, it is not at all clear why the primitive streak rather than, say, some later developmental markers should determine the beginning of individuality and humanity of the early embryo. In his article on the personhood of the human embryo, O'Rourke refers to three scientific facts enjoying wide recognition among prominent human embryologists. First, individual human life starts at fertilization when the sperm and ovum unite so as to form a one-celled zygote.[142] Second, "the zygote develops in an epigenetic manner," i.e., "through activation of potentiality" (O'Rourke 2006, 246), which means that "the structure and organs of the future fetus, infant, child, youth, and mature adult are present potentially in the one-celled zygote." Third, the zygote possesses "a genetic package (the human genome) that is the basis for future human development." In sum, "the zygote at the time of fertilization is not a potential human being; rather, it is a human being with potential" (O'Rourke 2006, 247). Thus, the denial of the human status of the early embryo before it reaches a certain point on the continuum of development, whether it be a primitive streak or, for example, brain organization sufficient for recording pain stimuli as others propose,[143] does not accord with relevant scientific data and as such cannot be morally supported.

According to the fourth and last argument provided by the *Encyclopedia of Bioethics*, it is the high ratio of natural losses of pre-implantation embryos that indicates their subhuman status and hence warrants their less-than-respectful treatment. One version of this position, put forth by Karl Rahner, argues that from a theological angle it is hardly acceptable that God gives souls to legions of embryos that die soon after (see Rahner 1972). In response, one must note that even if the mere fact of the destruction of early embryos appears uncontestable, it does not follow logically that these embryos lack individuality.[144] Furthermore, even if it seems implausible from a solely human standpoint that God would create a separate soul for each of these embryos, this neither negates the possibility of such creation nor validates disrespect toward the early embryos themselves. Admission of the boundaries of human knowledge is appropriate at this point as science does not demonstrate the capability of comprehensively explaining the phenomenon under review nor does the human intellect claim to be able to fathom God's plan with respect to it. What science does explain, however, is that implantation failures are not infrequently caused by chromosomal aberrations, possibly severe enough that an embryo never actually forms. With this being the case, the entities discarded before implantation may well not be embryos at all and hence may not receive souls from God (George and Tollefsen 2008, 137).

It is important to emphasize that the denial of the human moral status of early embryos results immediately in their abusive and unrestrained usage in research. Such embryos, treated as disposable material rather than as human beings, i.e., persons deserving of respect, cannot be called subjects *with whom* research is performed. Rather, being denied their own interests, they are viewed as the objects *on which* investigators experiment in the interest of others. Given that virtually all experimentation on pre-implantation embryos can not only be classified as nontherapeutic but also is lethal, no meaningful proxy consent can ever be secured for it. In summary, when the humanity of early embryos is admitted, proxy consent for nontherapeutic research is illicit.[145] If the humanity is denied, proxy consent is not even needed. In fact, although some sort of "consent" may be sought, for example, when permission is requested of parents for experimental use of so-called "spare embryos" formed in vitro, such "consent" contradicts the very purpose of informed consent not only by not safeguarding the human subject but also by sentencing it to death. What is more, assigning a proxy to decide on behalf of the early embryo raises serious questions because biological parents, even if available, cannot be assumed to be optimal advocates of the good of the embryo (cf. Harris, Morgan, and Ford 2004, 722). Roberto Colombo puts the matter well by stating that since "they are liable to the authority of others who may have interests conflicting with or neglecting the 'best interest' of the research subject," human embryos are "juridically vulnerable" (Colombo 2003, 226–27). From the biological point of view human embryos are the most vulnerable research subjects; they are *voiceless* in the fullest sense of the word. Furthermore, conceptual enfeeblement of their human subjectivity inevitably leads to the violation of their dignity and hence their basic rights as humans (see Colombo 2003, 214–15).

8.2.4.1 Embryonic Stem Cell Research

A telling example of the denial of the humanity of early embryos, or at the very least the rejection of their moral status as persons, i.e., as subjects of rights entitled to protection, is found in the heated debate over embryonic stem cell research, or more precisely, in the arguments of those who favor such research. To see how this is so, a brief explanation of what stem cells are will be helpful. According to the *Encyclopedia of Bioethics*, they are the cells equipped with "capacity to 'specialize' and become virtually any part of the organism into which they are introduced. Thus if introduced into the brain they become brain cells, if into the cardiovascular system they become cells of that type and so on" (Harris, Morgan, and Ford 2004, 722). "Pluripotent" stem cells develop in the human embryo very early after fertilization. They form the "inner cell mass" of the embryo during the blastocyst stage when the embryo is about to implant in the womb and then go on to form the hundreds of differentiated cells of the developing human being. Pluripotent stem cells have the capacity to form any of the more than 200 differentiated cells that make up the adult human body. In theory, if "extracted" early enough from the embryo (and this would require killing the embryo), they could be cultured and manipulated to become cells needed to effect cures of various diseases, e.g., Parkinson's disease (see Doerflinger 2004; Pellegrino 1999).[146]

As summarized by May, the exponents of embryonic stem cell research, although they focus primarily on the advantages of such research, nonetheless admit certain drawbacks. In particular, pluripotent stem cells are praised for their (1) flexibility, (2) capacity to "potentially provide an endless supply of cells," even from one stem-cell line and (3) availability, since they may be retrieved from so called "spare embryos" which are brought to life in vitro and put in cryopreservation. At the same time, however, these cells are (1) immunogenic (prone to rejection after transplantation); (2) tumorigenic (likely to produce tumors) and (3) "very difficult to differentiate uniformly into a target-type cell or tissue" (May 2000, 230). More important, though beyond the purely technical or scientific level at which these pros and cons are evaluated, a number of serious moral considerations arise.

Given that obtaining embryonic stem cells exerts lethal effect on embryos, the proponents of embryonic stem cell research have tried to circumvent its ethical implications by denying either the humanity or the personhood of the embryo. In these attempts, the personhood criterion has usually been brought up only to indicate that human embryos have no rights and hence can licitly be sacrificed in the interest of others. Yet, in line with May's objection, "this claim . . . drives a wedge between *being a human being* and *being a person*, and requires those who distinguish sharply between the two to offer a non-arbitrary criterion to distinguish which members of the species are persons and which are not" (May 2000, 230). Such a criterion has never been presented, however, and, indeed, as May concludes, it simply cannot be ascertained. The intentional slaying of human beings in the embryonic stage of development always remains intrinsically evil precisely because human subjects capable of full development into adult individuals are at stake. Moreover, two major ways in which embryos utilized in stem cell research are obtained, i.e., abortion and in vitro, are rightly considered as highly immoral.

All this, of course, is not to say that stem cell research ought to be ruled out as one of a range of morally justifiable options. A number of extra-embryonic sources of pluripotent stem cells have been identified. They may be drawn, for example, from bone marrow, placenta, or umbilical cord. Recently, scientists have recognized adult skin and amniotic fluid as very promising in this regard. It is quite significant that all these sources not only do not entail killing human embryos, but also are highly effective in treatments of numerous diseases.[147] In contrast, "up to now, no human being has ever been cured of a disease using embryonic stem cells" (Pacholczyk, On-line, accessed August 4, 2008).[148] Despite this advantage of adult stem cells, different academic, political, and media pundits have promoted embryo-destructive stem cell research and strongly demanded federal funding for its advancement. Adult stem cell research, on the other hand, is neglected in the mainstream agenda. Thus, the debate has become subject to ideological manipulation, with the powerful influence of political and financial interests instrumentalizing and objectifying to an even greater extent the most voiceless and powerless human beings.[149]

8.2.4.2 Genetic Therapy

The topic of genetic therapy belongs to a wide area of genetic engineering, the practice of which at a most rudimentary level denotes "the introduction of changes into the genes of a cell." In a broader sense, the term points to "the application of genetic sciences to cure disease or change human genetic endowment by altering the genes of a cell (Cochran 2007, 465)."[150]

As the *Encyclopedia of Bioethics* explains, two crucial distinctions organize the domain of genetic engineering. The first is the somatic versus germ line distinction, and the second differentiates between therapy and enhancement. Thus, genetic engineering breaks down into four types: "somatic gene therapy (SGT), somatic genetic enhancement (SGE), germ line gene therapy (GLGT) and germ line genetic enhancement (GLGE)" (Resnik 2004, 960).

Regarding the somatic vs. germ line distinction, the "somatic" signifies "genetic intervention pertaining to a specific disease in a specific person, altering that person's genetic code in order to treat or eliminate his or her disease." "Germ line," by contrast, refers to altering "the code in a germ cell so that the disease trait would not be passed on to the patient's descendants" (Cochran 2007, 465). Put another way, "if somatic genetic therapies are to genetic disease what antibiotics are to infectious disease, then germ line genetic therapies are the equivalent of vaccinations—prevention rather than recovery" (Prentice 2001, 534).

When it comes to the difference between therapy and enhancement, the latter's goal, in opposition to the former, "is not to treat or prevent a disease but to achieve another result, such as increased height, intelligence, disease resistance, or musical ability" (Resnik 2004, 961). Not surprisingly, then genetic enhancement in both somatic and germ line forms gives rise to serious moral objections, as it expands the traditional goals of medicine and, in the eyes of many, extends far beyond its legitimate boundaries. Moreover, it is far from clear, to say the least, that people possess the moral mandate to retrofit themselves with more efficient genetic material.[151] From the theological standpoint, genetic enhancement leads to the rejection of human life and nature as God's gifts, because it seeks complete control over one's physical and psychological performance.[152] One example of this may be the denial of vulnerability inherent to the human condition, combined with a determined endeavor to eliminate suffering. Thus, the focus shifts from alleviating suffering, one of medicine's long-standing and widely recognized tasks, to eradicating all its forms from human experience (O'Mathúna 2002, 294). Moreover, at a philosophical or even at a common-sense level, one realizes that besides its beneficial potential, genetic engineering "also represents the darker side of human freedom and creativity—the denial of death and of human finitude, the lure of profit and power, and the exclusion of the disadvantaged" (Cochran 2007, 465). Finally, genetic enhancement technologies inevitably become entangled in financial and marketing issues as well as in social inequalities.

Consequently, it is not hard to envision that "persons with power and money will demand the best for their children, commanding in the future access to physical and mental enhancements through genetic medicine just as they demand today access to the best schools and health care." Obviously, "those without power and resources will lose in the competition, raising the specter of a new form of eugenics" (Cochran 2007, 465).

Regarding the morality of somatic gene therapy, it is important to note, along the lines of the summary provided by May, that its development has not yet reached beyond an experimental stage, which is to say that it cannot be considered standard treatment and as such holds the potential for unknown risks and side effects. As is the case with other experimental treatments, one should resort to somatic gene therapy only when no established alternative procedures can yield satisfactory therapeutic results. A highly favorable proportion of known risks and expected benefits must be assured in order for somatic gene therapy to attain the status of moral permissibility (May 2000, 240).

Although, like its somatic counterpart, germ line gene therapy is considered as purely experimental, other issues make it the subject of an even more detailed scrutiny. As alluded to above, such therapy can be compared to traditional preventive treatments. This comparison limps, however, because germ line gene therapy "affects *prospective* individuals, and not just the one individual, but all of the generations of progeny from that individual" (Prentice 2001, 534). Here a perplexing question immediately arises: does anyone have the right thus to impact his or her descendants, and if so, on what grounds? Even if, as often claimed, such an impact proves beneficial to human beings, who has the authority to decide in what such improvement should consist? Put otherwise, what truly makes humans better and what does not?

Related to these questions are the following three quandaries, brought up by David Prentice. First, germ line genetic engineering is both unpredictable and uncontrollable. To give but one example, neutralizing a gene responsible for a malady may yield dangerous unintended effects, as certain genes, while conducive to one disease, at the same time safeguard against another. Such is the case of "the mutation which causes sickle cell anemia" that "actually has a protective effect against malaria." Therefore, it is impossible to foretell "the consequences of alterations to the germ line" and the results of the bequeathal "of that new 'version' of the human genome to future generations." Second, germ line genetic procedures will inevitably engender "a genetic caste system," the corollary of a societal breakup into different classes of citizens separating those of "manufactured 'genetics'" from "the 'normals,'" or even "into different species if the modifications are extensive enough" (Prentice 2001, 538), This, in turn, will only exacerbate the economic and political disparities with which the world is already rife, by adding a new form of elite genetically redesigned aristocracy likely to prune away those "not-improved." Third, in some of its forms, germ line genetic engineering results in the treatment of some people like chattel, as, for example, when they are programmed in advance with the sex and the physical traits deemed advantageous by those in power. Put another way, such engineering sooner or later turns into the manufacture of human beings, which grossly offends their dignity and inevitably leads to treating some human beings as property of others "designed and crafted to meet the maker's desires." If this kind of engineering prevails, man will end up "making man in his own image, yet without any higher standard to which the crafts-man is held" (Prentice 2001, 539).

Even so, it is important to emphasize that in the course of time some kinds of germ line therapy may become more resistant to all the charges leveled above. In its document "Communion and Stewardship," the International Theological Commission admitted that such therapy "would in itself be acceptable were it not for the fact that is it is hard to imagine how this could be achieved without disproportionate risks especially in the first experimental stage, such as the huge loss of embryos and the incidence of mishaps, and without the use of reproductive techniques." Still, "a possible alternative would be the use of gene therapy in the stem cells that produce a man's sperm, whereby he can beget healthy offspring with his own seed by means of the conjugal act" (n. 90). In the same vein, May submits that while "at present such 'therapy' seems unrealizable, and the risks entailed . . . are far too great and even unknown, to warrant its use," it certainly is not intrinsically evil, and "it might perhaps be morally licit in the future" (May 2000, 241).

8.2.4.3 Prenatal and Pre-implantation Screening

In recognition of the substantial benefits that may be brought about by prenatal screening, the Magisterium approves it as morally justifiable, subject to several qualifications.[153] In line with John Paul II's encyclical letter *Evangelium Vitae*, prenatal diagnosis, "if carried out in order to identify the medical treatment which may be needed by the child in the womb"[154] is morally licit (n. 14), yet, as the Instruction *Donum Vitae* specifies, the following additional requirements must be met: (1) informed consent from parents is to be procured after providing them with adequate information; (2) the methods used "safeguard the life and integrity of the embryo and the mother," and (3) neither the embryo nor the mother is put to disproportionate risks (part I, n. 2).[155]

A serious moral problem arises when prenatal screening, as John Paul II put it, "becomes an opportunity for proposing and procuring an abortion" (John Paul II 1995, n. 14). The Pope condemns such abortion and calls it eugenic. By the same token, *Donum Vitae* points out that "a diagnosis which shows the existence of a malformation or a hereditary illness must not be the equivalent of a death-sentence." Therefore, "a woman would be committing a gravely illicit act if she were to request such a diagnosis with the deliberate intention of having an abortion should the results confirm the existence of a malformation or abnormality" (part 1, n. 2).[156] The importance of intention cannot be overestimated in the present-day context, when the majority of prenatal tests are aimed to terminate pregnancy as soon as a serious anomaly, such as, for instance, Down syndrome, Trisomy 18 or 13, or spina bifida is detected. Moreover, within the United States, the law commonly requires physicians to offer such tests and the failure to do so may result in malpractice suits (May 2000, 242).

The problem of the intentional killing of the unborn who suffer from disease also applies to pre-implantation screenings,[157] typically "performed on a single cell from an eight-cell stage embryo" (Byrnes 2001, 559). Such screenings originated in the wake of in vitro technology, as of today the primary source of embryos to be used for screening purposes.[158] Obviously, if used with the intention of exterminating the embryo if it turns out to carry serious diseases, pre-implantation diagnosis does not live up to moral standards. Moreover, this immorality is aggravated insofar as human embryos are brought to life through in vitro fertilization (IVF). The Magisterium firmly states that IVF never presents a morally valid option.[159] When free from these issues, however, pre-implantation diagnoses may prove beneficial and morally permissible (May 2000, 247). Nowadays, Ashley and O'Rourke note, "there are approximately four thousand different diseases in the single gene disorder group, all of which are caused by different abnormal genes." In addition, "33 percent of infant deaths are related to genetic causes" (Ashley, DeBlois and O'Rourke 2006, 99). Even though therapeutic interventions are still largely under development, the future may bring solutions that will obviate both technical and ethical obstacles and thus render pre-implantation screening both medically advantageous and morally licit.

Closely related to both pre-implantation and prenatal diagnoses is the topic of genetic counseling which has as its purpose to aid individual persons, couples, or families in checking for and responsibly facing the prospect of genetic disease.[160] Couples typically seek such counseling "because of fears about possible defects in children already in existence or about their responsibilities for future pregnancies" (Ashley, DeBlois, and O'Rourke 2006, 101). *The Ethical and Religious Directives* approves such efforts insofar as they are integrated with responsible parenthood and aim at "the proper treatment and care of children with genetic defects, in accordance with Catholic moral teaching and the intrinsic rights and obligations of married couples regarding the transmission of life" (n. 54). Whenever evidence exists that the couple can, indeed, conceive a child with genetic abnormalities, such means as contraception, abortion, or in vitro fertilization, because of their gravely immoral character, must never be advised. From this it does not follow, however, that the couple has an obligation to take the risk inherent to conjugal acts that may result in the conception of a child with a genetic disease. In his encyclical letter on the regulation of birth, *Humanae Vitae*, Pope Paul VI contends that responsible parenthood is exerted not only "by those who prudently and generously decide to have more children" but also "by those who, for serious reasons and with due respect to moral precepts, decide not to have additional children for either a certain or an indefinite period of time" (n. 10).[161] No doubt, "serious reasons" would include the possibility of a serious genetic malady, especially when it cannot be treated by genetic therapy. As commented upon by May, it would not be immoral for the couple to conclude that "running the risk would either be unfair to others or that they either could not fulfill or would be seriously tempted to omit fulfilling the responsibilities they would incur in caring for a child afflicted by the malady." On the other hand, however, the couple may also "courageously agree to accept the risk and by doing so firmly commit themselves to carry out the responsibilities that they would incur should their child indeed be afflicted by the malady in question" (May 2000, 248).

Thus, the decision is ultimately left to prudent consideration of the parents with the presumption that they act in harmony with natural moral law and the obligations it entails.

8.3 Conclusion

In this chapter, I have provided the foundation for considering the issue of proxy consent in nontherapeutic research. Whether or not proxy consent is justifiable in this setting is the topic to be taken up in the next chapter. Here, the positions of Paul Ramsey and of the earlier William May that deny such justifiability will be examined. For the purpose of clarity the development of their views will be contrasted with the parallel development of the thought of Richard McCormick. Then the exposition of the debate will continue with a primary focus on the arguments supporting the legitimacy of proxy consent in the nontherapeutic milieu.

Chapter 9
Proxy Consent in Nontherapeutic Settings

Two major lines of argumentation against the justifiability of proxy consent for nontherapeutic research offered by Ramsey and the earlier May unfolded dialectically in the course of their debate with Richard McCormick, and so it seems appropriate to place the latter's most relevant views on the matter here, even though technically they ought to be classified as part of the following chapter, which examines arguments that favor the validity of proxy consent in nontherapeutic situations. What is known as the Ramsey and May vs. McCormick debate began in the 1970s and even though its content centered largely on children subjected to nontherapeutic research, it can also be applied by extension to other classes of incompetent subjects. Despite the numerous ethical insights that they shared and the support they offered one another in the exchange against McCormick, Ramsey and May developed their own approaches, which are not entirely equivalent. For the sake of clarity I have decided to present them separately and in contrast to McCormick's position.

9.1 Paul Ramsey vs. Richard A. McCormick

The content of this subsection derives from and follows on the concise introduction to Ramsey's views on informed consent offered previously in this study. Certainly, the way in which Ramsey accounts for proxy consent is consistent with his convictions on informed consent in general. As he puts it, "from consent as a canon of loyalty in medical practice it follows that children, who cannot give a mature and informed consent, or adult incompetents, should not be made the subjects of medical experimentation" (Ramsey 2002, 11). This assertion notwithstanding, Ramsey allows for therapeutic research with the incompetent under two conditions: (1) when no panacea is available and (2) when "it is reasonable to believe that the administration of a drug as yet untested or insufficiently tested on human beings, or the performance of an untried operation, may further *the patient's own recovery*" (Ramsey 2002, 12, emphasis in text).

Nontherapeutic research on minor subjects, however, must be disapproved unconditionally, according to Ramsey, because it violates faithfulness-claims that

G. Mazur, O.P., *Informed Consent, Proxy Consent, and Catholic Bioethics*, Philosophy and Medicine 112, DOI 10.1007/978-94-007-2196-8_9,

the child puts upon the investigator and indeed on everyone by virtue of his or her humanity. Faithfulness, which embodies the canons of mutual loyalty, infringes upon the discretion of parents or legal guardians in regard to proxy consent for subjecting the child for whom they are responsible to experiments. Ramsey states that "no parent is morally competent to consent" (Ramsey 2002, 13) for his or her child in the case of nontherapeutic research. Therefore, neither parent nor guardian should ever be asked to consent to such research (Ramsey 2002, 14).

Ramsey's position appears absolutist: the mere expectation of relaxing the moral ban on nontherapeutic research involving children forestalls faithfulness and loyalty toward them. The fundamental assumption here is that for nontherapeutic research to reach the status of moral soundness it must be rooted in genuine consent, and that is exactly what children and other incompetents are not capable of giving. Only by free consent may competent adults volunteer to participate in research unrelated to improvement of their own condition. Volunteering the incompetent for such research constitutes an infraction of the canon of loyalty and hence undermines the role that parents or other proxies ought to play in regard to their children or wards (Ramsey 2002, 14, 24). Perceived as embedded in "the covenantal obligations" (Ramsey 2002, 36) and expressive of "the highest fiduciary loyalty" aiming at the protection of the child, parental care necessarily implies fidelity (Ramsey 2002, 25). Fidelity to the child, Ramsey explains, in recourse to Kantian terms, "includes never treating him as a means only, but always also as an end" (Ramsey 2002, 35). I will return to this formulation below in the context of McCormick's critique of Ramsey's argumentation.

It is noteworthy that Ramsey opposes nontherapeutic research with the incompetent as subjects regardless of the degree of risk involved. Consequently, he excludes from the scope of morally permissible options even those procedures that feature no more than minimal risk. In this regard, two factors stand out. First, determining what amount of risk precisely counts as minimal is far from easy and prone to abuse (see e.g. Ramsey 1976, 26–29). Second, regardless of the risk, "the subject can be wronged without being harmed" (Ramsey 2002, 39). With respect to the latter, Ramsey employs the notion of "unconsented touching" (Ramsey 2002, 37). In brief, even with a hypothetical zero risk involved in experimental procedures, the child may still be touched offensively, i.e., without informed consent, which cannot be validly substituted by proxies except for therapeutic research. Unconsented touching objectifies the child even when no risk is involved. As a matter of fact, however, zero-risk situations in research virtually never occur. What is more, not only is minimal risk at issue in nontherapeutic research but also, as Ramsey explicitly admits, "unknown and undiscerned risks" often exist "in the case of an invasion of the fortress of the body". This in itself constitutes just one more reason, albeit a significant one, for holding that procedures that hold out no benefit, however remote, to the incompetent subject "can be consented to by an adult in behalf of a child *only* if it is in the child's behalf medically" (Ramsey 2002, 52, emphasis added).

To reinforce this conclusion, Ramsey argues that an attempt to consent to the subjection of a child to nontherapeutic research essentially means "to treat a child as not a child. It is to treat him as if he were an adult person who has consented to become a joint adventurer in the common cause of medical research." This may be done on the basis of "presumptive or implied consent of the child" but such presumption is always "violent and false" (Ramsey 2002, 14). It is here that the difference between the positions of Ramsey and McCormick forcefully arises. In McCormick's view, proxy consent becomes "morally valid precisely insofar as it is a *reasonable presumption* of the child's wishes, a construction of what the child would wish could he consent for himself" (emphasis added). In order to avoid arbitrariness or "a simple imposition of the adult world on the child," McCormick endeavors to ground his approach in an objective understanding of moral duties based on natural law, and adds that the child "would choose ... because he *ought* to do so" (McCormick 1981b, 61),[162] To this Ramsey counters that there is no need to resort to natural human inclinations as providing the basis for the moral "ought" in proxy decision making because the only good to which childhood inclines is "the preservation of life and a healthy life and growth ... and that constitutes parental care in the early years" (Ramsey 1976, 22). By seeking to embed presumptive consent in other goods to which the child ought to consent because we all ought to do so (McCormick 1975, 128), McCormick in fact imposes the adult world on the child, Ramsey states. What is more, the balance of McCormick's argument provides even clearer evidence in this regard.

After arguing that proxy consent in therapeutic research "is morally legitimate ... precisely because we know that life and health are goods for the child, that he *would* choose them because he *ought* to" (McCormick 1981b, 61), McCormick goes on to ask whether there are "other things that the child *ought*, as a human being, to choose precisely because and insofar as they are goods definitive of his growth and flourishing?" (McCormick 1981b, 62). In particular, ought the child to participate in nontherapeutic research and if so, under what conditions? In answer, McCormick first notes that from the fact that all human beings are social it follows that we should not only pursue the good of our own lives but also, whenever possible, the good of the lives of others. In fact, "the goods that define our growth and invite to it are goods that reside also in others" and it can be good for us to realize and endorse those goods not only in ourselves but also in others. What is more, under certain circumstances, we ought to pursue them. And "if this is true of all of us up to a point and within limits," McCormick continues, "it is no less true of the infant" (McCormick 1981b, 62). Consequently, subjecting a child to nontherapeutic experimentation can be valid "up to a point and within limits," precisely because one can reasonably assume that the child would if he could—because he ought to—contribute to the good of medical science or pursue the goods of other people who in the future may benefit from the results of the experiment.

With this in mind, McCormick moves on to determine the meaning of the phrase "up to a point," through which he attempts to define the obligation of sharing and of pursuing the good of others. In doing so, he begins by making two relevant statements. First, whether or not participation in nontherapeutic research proves good for the incompetent subject is "a very circumstantial and therefore highly individual affair" (McCormick 1981b, 63). Second and more fundamentally, such participation becomes morally good only insofar as it is undertaken voluntarily. It follows that no third parties are equipped with a moral mandate to presume the consent on behalf of the subject. Yet, this general rule notwithstanding, in certain highly specific circumstances one can reasonably assume that everyone would wish to participate in a given research project because everyone ought to. For example, "when a particular experiment would involve no discernible risk, no notable pain, no notable inconvenience, and yet hold promise of considerable benefit" the child should "be constructed to wish this in the same way we presume he chooses his own life, because he *ought* to". In McCormick's own terms, the child, "*ought* to want this not because it is in any way for his own medical good, but because it is not in any realistic way to his harm, and represents a potentially great benefit to others" (McCormick 1981b, 64).

In further elaboration on this point, McCormick refers to two levels at which good can be pursued: (1) an extraordinary level which is supererogatory and highly individualized; this is the realm of charity in which no one can reasonably presume for another that he *ought to* pursue the good involved; (2) an ordinary level at which realizing the good constitutes moral obligation for everyone who faces such a possibility. This is, McCormick states, the domain of justice, and even infants, by virtue of their sharing in human social nature, can be reasonably assumed to pursue the good contained in it. The ordinary level involves no discernible risks, costs, or other inconveniences for the subjects but it does involve substantial benefit for others. *Up to the point* that the nontherapeutic research meets the requirements of the ordinary level, McCormick asserts, the incompetent's wishes to participate can be reasonably presumed on the basis that they "ought to" consent to such participation.

Ramsey admitted that McCormick's ceiling for enrollment of children for nontherapeutic research was set very low but nonetheless found its rationale unpersuasive. The child's sociality undergirding the presumption of consent may include only a fairly narrow scope of nontherapeutic situations. Nevertheless, the ordinary level of moral "ought" introduces another threat: easy justification of potential and involuntary utilization in nontherapeutic experiments of other, non-minor, subjects "provided only that they are needed and that the risks are minimal" (Ramsey 1976, 24). Moreover, the criterion of minimal risk itself proves insufficiently specific and thus far from satisfactory.

Similarly, Ramsey charges McCormick with creating an "accordion morality" that consists in a set of expressions such as "no discernible risk," "no notable inconvenience," "almost no cost" but with no clear meaning attached to any of these expressions. As a matter of fact, McCormick vacillates among these and similar terms by which he fails to provide a reliable moral yardstick for valid enrollment of minors in research. His approach, according to Ramsey, resembles "an accordion that can be expanded and contracted according to the need for results." To remedy this failure, Ramsey proposes an attempt to "close the gap" between significant and insignificant risks by "a further discussion of cases, without usurping the proper judgment of physicians" (Ramsey 1976, 27).

From what has been said thus far it follows that Ramsey's criticism of McCormick's justification of proxy consent in nontherapeutic situations comes down to two major charges: (1) McCormick treats children as small adults by imputing moral obligations/moral agency to them, a rendering that also opens the door to involuntary use of research subjects other than minors; (2) McCormick's all-too-flexible approach yields convenient solutions dependent on the needs being addressed.

The issue of moral agency presented in the first charge was originally brought up and articulated thoroughly by May and thus will be discussed at greater length in the following section. Nevertheless, Ramsey also made this an essential part of his own account and his work was repudiated by McCormick on the grounds that moral responsibility is not so much rooted in adulthood as it is in human sociality. McCormick explains this as follows: "When we say of adults (who can and do have obligations) that it is reasonable to expect that they will want certain goods for others and contribute to these goods if there is no discernible risk, discomfort, or inconvenience, it is not precisely because they are adults that we conclude this, but *because they are social human beings*" (emphasis in text). In other words, the "ought" language with respect to children does not hint that they are moral agents. Rather, "it is simply a device, a construction (as is also the language of what the infant *would* choose) to get at the reasonableness of our expectations and interventions" (McCormick 1981c, 90).[163] Therefore, "in using the language of *ought* we concentrate on sociality, not age. And in doing so . . . we introduce the infant not to the world of adults, but to the world of *social* human beings, to the human world" (McCormick 1981c, 91).

It is true, McCormick concedes, that one can expand the implications of the "ought" approach onto other groups of subjects, but this in no way constitutes an objection. On the contrary, if, for instance, "not enough volunteers are available for minimal-risk experimentation and the research seems of overriding importance to the public health, it would not be unjust of the government to recruit experimental subjects, for example by lottery, just as it is not unjust for government to draft soldier for national self-defense" (McCormick 1981c, 92). The fact that Ramsey protests against the reasoning that may lead to such an inference is evidence, according to McCormick, of his "narrowly individualistic notion of human nature" (McCormick 1981c, 91).

McCormick finds support for this observation in two other characteristics of Ramsey's thought. The first of these characteristics is individualism in addressing some individual rights, an individualism that betrays no connection to "needs and claims of others." Yet, "in saying that *any* nontherapeutic experimentation on an infant is violative of the infant's right, Ramsey must suppose that there is an established right precluding this—the very point to be proved." This point can be proved only, McCormick maintains, "by a very one-sidedly individualistic notion of rights and claims" (McCormick 1981c, 93). The second dubious characteristic of Ramsey's thought is that it builds upon the Kantian dictum that forbids treating the child "as a means only," but rather "always also as an end" (Ramsey 2002, 35). The words "only" and "also" used here merit greater attention. If, for example, a competent woman agrees via informed consent to participate in research, she becomes *not only* an object of research *but also* and indeed in the primary sense a subject. In Ramsey's language, she is a subject who does something, not just an object "done to" (Ramsey 1976, 26). Clearly, she first *does* consent to research and then an experimental procedure is performed *on* her. At one level, by subjecting herself to research the woman treats herself as a means, but certainly not as a means only. Besides being an object experimented *on*, she is a subject *with* whom research is conducted.

Here, however, lies a major difficulty with children, who cannot by virtue of free, informed consent render themselves subjects of research. If they are volunteered by others they are thereby made objects, and unless the research is therapeutic, they are objects only. In other words, in nontherapeutic research, when no direct link exists to their own medical needs, children are used as mere means and therefore they are abused in the interest of and to the benefit of others (Ramsey 2002, 12, 36). Accordingly, Ramsey submits that "research possibly or even remotely connected with treatment is the *outer limit* of procedures that treat a child-patient as an 'end *also*' and never as a 'means *only*'" (Ramsey 1976, 26).

This, in turn, implies a broad definition of therapeutic research, in which not only does a proximate medical advantage play a vital part but a remote benefit also enters into the equation. In therapeutic research, the child ceases to be the mere means to scientific ends because his own welfare (ends, interests) are taken into account seriously enough so as to take precedence over the mere realization of scientific values or the interests of other parties.[164] While McCormick tolerates the Kantian maxim, he nonetheless resists the way in which Ramsey applies it to the point in question, stating that "if the individual's good transcends individuality and separateness from others, then to treat him or her with this in mind need not involve treating as a means *only*." More relevantly, one needs to "establish the limits of the implications of our sociality once insertion into sociality has been done" (McCormick 1981c, 94). Ramsey's individualistic, asocial take on children led him to assume, from McCormick's standpoint, that proxy consent for nontherapeutic research must treat the incompetent subject as a means only and thus he failed to consider the social dimension of human existence as possibly providing a basis for justifying a different conclusion.

The individualistic perspective permeating Ramsey's account on proxy decision making contributes, in McCormick's eyes, to the generation of fears regarding the employment of the incompetent in nontherapeutic research. Ramsey enunciated one of these fears in relation to the minimal-risk criterion on the basis of which he then leveled the "accordion morality" charge against McCormick. McCormick, in turn, defended his original position by making two points. First, given their very nature, moral judgments cannot be formulated with mathematical precision and hence they allow for some degree of elasticity. This is clearly seen in the case of minimal-risk judgments in which ethical assessments require consideration of scientific data that itself may not be exact. Therefore, and this is McCormick's second point, it is the scientific community, not moral theologians, who are in a position to determine which medical procedures carry no more than minimal risk and which exceed this threshold. Even if Ramsey seems to agree with this, he nonetheless calls for mathematical precision in a domain where both moral judgments and also the scientific information on which they rest are indistinct and obscure (McCormick 1981c, 96–97).

9.2 The Early William E. May vs. Richard A. McCormick

Quoted by Ramsey and closely aligned with him during the debate against Richard McCormick, May worked out his own position, which featured as its basic tenet that "every human being, by being a member of the human species, is a being of moral worth, with rights that demand recognition, respect and protection" (May 1977, 21). Moral worth implies the existence of an intrinsic and inalienable dignity by virtue of which all humans are "equal in their humanity" and as such they cannot licitly be subordinated to the interests of other individuals or society.

In the realm of human experimentation it is informed consent that guards the moral worth of research subjects. Consequently, "to experiment on a human subject without securing his consent is to treat him as a being who is no longer of moral worth; to make of him a means, not an end; to subordinate him to other humans; to repudiate his humanity" (May 1977, 21). Yet, in the case of those who because of compromised competence cannot offer valid informed consent, can proxy consent be secured on their behalf? After noting that broad agreement exists among scholars in regard to the permissibility of such consent in therapeutic settings, May admits that much less unanimity can be found as far as nontherapeutic research is concerned. Having found himself in substantial agreement with Ramsey, May thoroughly examines McCormick's position and rejects it on the grounds now to be presented.

At the outset May points out that McCormick's legitimization of proxy consent in nontherapeutic research depends on the justification of proxy consent that he offers for therapeutic settings. After examining the latter and finding it incorrect, May infers that McCormick, a fortiori, errs also on the former. But why precisely are McCormick's attempts to vindicate the validity of proxy consent in therapeutic research flawed? In agreement with Ramsey, who believed that the very term "consent" "is to some degree false" (Ramsey 2002, 14, note 11) when used in reference to permission given by proxy for the subjection of another person to research, May notes that all the incompetent persons "for whom 'proxy consent' is offered have in common two supremely important characteristics or features. These are (1) that they are all *beings of moral worth*, that is entities who are the subjects of rights that transcend the societies in which they live and that must be recognized and respected by their fellow men, and (2) that they are *not moral agents*, that is entities who are the bearers of moral obligations or duties" (May 1974, 247). Strictly speaking, then, the term "proxy consent" may be puzzling as it in fact denotes not the presumed or constructed consent of the subject but rather the personal consent of a proxy given on behalf of the subject (May 2007, 246). In other words, the term makes for "a false presumption or construction" because "consent is a *human* activity; it is an act that requires knowledge and freedom of choice in order to exist. It is, in other words, a *moral* act, and as such it can only issue from a *moral agent*" (May 1974, 247). Consequently, McCormick's account of proxy consent cannot be condoned because it postulates that minors and other incompetent are moral agents, the carriers of moral duties.

This obviously does not tally with reality and prompts May not only to reject McCormick's argument but also to search for another sound rationale for proxy decision making in therapeutic research. Such a rationale, according to May, "is not to be found in any presumed duties or obligations attributable to the subject in whose behalf consent is given. Rather, it is to be sought in the duties or obligations that do, in truth, relate other members of the human community to that subject" (May 1974, 248). To explain the meaning of the latter statement, May refers to the "Kew Gardens Principle" first introduced by John G. Simon, Charles W. Powers, and Jon P. Gunnemann[165] in order to provide grounds for the moral responsibility one bears when witnessing injustice done to another person by other human beings. Four factors are relevant here: (1) the need: a person wronged by another is in need of help because some good, e.g., the good of life or of physical integrity, is being menaced or eliminated; (2) proximity: one is conscious of another person being in need of help; (3) capability: one is capable of coming to another's aid; (4) "last resort": although not clearly defined by the above mentioned authors, this requirement is met when all three of the first conditions are met.

The application of the "Kew Gardens Principle" in proxy consent to therapeutic research with a child as subject can be performed in the following way. First of all, it is clear that the child is in need. The child needs therapy (in this case an experimental one) because the good of his health or perhaps even his life is imperiled. The parents or legal guardians acting as proxies for the child are certainly conscious of the child's need. They are also capable of responding to this need and decide for the child, a very condition of acting as proxy. In May's own words, in such a situation "it would be irresponsible, immoral for parents and others *not* to take effective steps." The parents' consent is truly theirs, and "it is simply erroneous to speak meaningfully at all about any consent on the part of the child or infant or other human being who simply is not a moral agent" (May 1974, 249–50).

It is important to note, however, that the "Kew Gardens Principle" does not apply in nontherapeutic research. The child is in no need of help at all (medically speaking) which is to say that none of the goods belonging to him, such as life, health, or bodily integrity are in jeopardy. For this reason, the proximity and capability conditions become meaningless and hence the entire principle goes awry. In addition, May goes on to repeat that, because the child is not a moral agent, he has no obligation whatsoever to contribute to the common good or the benefit of others, and hence his possible enrollment in nontherapeutic research cannot be warranted on the basis of such an alleged obligation. Still, "the child or other incompetent human subject *is* a being of moral worth, a subject of inviolable rights, and adult members of the human community who *are* moral agents and bearers of moral obligations must respect him as such" (May 1977, 28). From this standpoint, to subject the child to nontherapeutic experimentation denies this respect because it mistakes the child for someone he is not, that is, an adult equipped with moral agency, the bearer of moral responsibilities and duties.

McCormick's reply to May's objections is twofold. First, he defends his position on the grounds that it need not "necessarily regard the infant as a moral agent. Nor need it imply that he has obligations. It need only suggest that what it is reasonable and legitimate to do experimentally with youngsters might be constructed off what others who are moral agents *ought* as humans to do; for though they are not yet moral agents, infants are humans in the fullest sense." Second, McCormick points to the fact that both Ramsey's and May's accounts unconditionally shut the door on proxy consent in nontherapeutic research, which eliminates from among the morally licit options even the most routine nontherapeutic procedures. This, McCormick believes, is too restrictive an approach and indeed it is one that is seen by numerous scholars "as unreasonable and at variance with common sense" (McCormick 1975, 128).

May finds the argument presented by McCormick less than satisfactory, and counters it with a quotation from McCormick himself showing that he "not only implied that the infant has moral obligations but *asserted explicitly* that this was the case" (May 1977, 29). Furthermore, according to May, although it is true that minors and the incompetent "are humans in the fullest sense" this does not render them automatically the carriers of moral obligations. Every child is only "radically capable of *becoming* a moral agent" and as a social being he needs the aid of the community for this end to be achieved. Certainly, however, the mere fact of sharing in humanity does not suffice to imply that one ought to do something in the moral sense.

Even if correct in his recognition of the "corporate, social character of our existence as human beings," McCormick nonetheless does not appreciate, in May's view, that there is "a vast difference between the way this dimension . . . is expressed in a child and in an adult." A child, as opposed to an adult, "is as yet not aware of his social obligations. His sociality consists in his need and right to be cared for by others and to arrive, with their help, at an understanding of himself and his responsibilities as a social being. But until he arrives at this understanding he is incapable of contributing to the common good by *his actions*, inasmuch as he simply cannot engage in human actions as a child . . . As a child he contributes to the common good just by being" (May 1977, 33). May adds further that submitting a child to nontherapeutic research "ruptures the covenantal bonds that ought to exist in and among human beings" as it considers "this human subject, this being of moral worth, either as an impersonal 'it' or as a bearer of moral obligations, neither of which he is" (May 1977, 31). To put it another way, because "human existence is *covenantal in character*" (May 1977, 30) any account of human sociality or social solidarity that aspires to reliability and soundness must give sufficient weight to this truth.

Finally, in response to McCormick's second point, which judges an unreserved prohibition of proxy consent in nontherapeutic experimentation to be too rigorous and reasonless, May claims that a breach of the covenantal bonds that exist among humans can come about in research regardless of the insignificance of the procedures involved. Indeed, offensive and objectifying use of another as an experimental subject does not have to include any substantial risks. What is more, "to touch human beings offensively by 'volunteering' them for an adventure in which they cannot, by reason of their condition, be volunteers is to attack not simply them but the dignity of all human beings" (May 1977, 33).

Having examined the arguments of Ramsey, the early May, and McCormick, I will now consider further developments and arguments regarding proxy consent in nontherapeutic research. The following chapter provides a critical revision of some premises endorsed in the course of the debate and then turns to the most recent arguments on the matter, and seeks to lay the foundation for a sound ethical and anthropological rationale, justifying the qualified permissibility of employing incompetent subjects in experimentation that, while holding out no immediate benefit to them, redounds to the good of others.

9.3 The Evolution of the Debate

Both Ramsey's and McCormick's analyses of proxy consent in nontherapeutic setting have received a great deal of attention from the scholarly community and have also stimulated and informed further debate on this controversial matter. In what follows, the four major positions advanced during the debate to justify permitting the participation of minors in minimal-risk experiments that do not benefit them medically will be laid out. The position examined first, offered by the National Commission for the Protection of Human Subjects of Biomedical and Behavioral Research, has the additional merit of critically summarizing not only the remaining three positions, but also those of Ramsey and McCormick. While elaborating on all of these arguments I will include the Commission's insights and remarks, by virtue of which the Commission's own stand will also, one may hope, become better understood and evaluated.

9.3.1 The National Commission for the Protection of Human Subjects of Biomedical and Behavioral Research

In its report *Research Involving Children*, issued a year before the Belmont Report, the National Commission took a stand on proxy consent less stringent than the positions of Ramsey, May, and McCormick. After recognizing that "research involving children is important for the health and well-being of all children and can be conducted in an ethical manner" (Recommendation 1)[166] the Commission recommended that research that does not carry with itself more than minimal risk[167] be performed on minors, subject to IRB approval. In addition, appropriate steps ought to be taken in order to secure the assent of the child (unless unfeasible) along with the authorization of the parents or legal guardians (Recommendation 3).[168] Like Ramsey and May, the Report stresses the difference between consent as an instance of one's autonomous action on the one hand and permission offered on behalf of someone else on the other. In precise terms, proxy consent refers to the latter notion rather than to the former. The permission of parents or legal guardians still differs from the assent of the children, which must be obtained when they have reached the age of seven (comment on Recommendation 7).

Regarding therapeutic experiments involving more than minimal risk, the Commission finds them justifiable provided that the risk-benefit ratio they entail "is at least as favorable to the subjects as that presented by available alternative approaches" (Recommendation 4). Remarkably, the Commission also proposes, albeit not unanimously,[169] that nontherapeutic research featuring risk greater than minimal may also be approved, but only when (1) accretion over minimal risk is slight, (2) the experimental procedure itself "presents experiences to subjects that are reasonably commensurate with those inherent in their actual or expected medical, psychological or social situations, and is likely to yield generalizable knowledge about the subjects' disorder or condition." Furthermore, (3) "the anticipated knowledge" must be "of vital importance for understanding or amelioration of the subject's disorder or condition" (Recommendation 5).

Commenting further on these requirements, the Commission explains that reasonable commensurability represents similarity and indicates "the procedures that the prospective subjects and others with the specific disorder or condition ordinarily experience (by virtue of having or being treated for that disorder or condition" (comment on Recommendation 5). Strikingly, no mention is made of healthy subjects to be enrolled in research including slightly more than minimal risk, and the wording employed in the passages just invoked clearly states that such research ought to be connected, even if remotely, to the subject's actual medical condition. By specifying within the same comment that "the research [in question] must hold out the promise of significant benefit in the future to children suffering from or at risk for the disorder or condition (including possibly the subjects themselves)," the Commission strives to set up a boundary for justifiability of nontherapeutic research with slightly-more-than-minimal risk involved. In fact, referring to such research as nontherapeutic also implies a stricter notion of therapeutic research, a considerably narrower one than that adopted by Ramsey, for instance. As already mentioned, for Ramsey, any connection, however distant, to the subject's actual condition suffices for a research project to be classified as therapeutic even though without immediate benefit to the subject.

Despite its efforts to limit unjustifiable enrollment of children in nontherapeutic research and to circumscribe the scope of such research, the Commission nonetheless leaves the options open as it allows for overriding virtually all its recommendations, except for the preliminary requirements listed in Recommendation 2 along with those concerning the child's assent and parental permission, provided that (1) "research presents an opportunity to understand, prevent or alleviate a serious problem affecting the health or welfare of children" and (2) "the conduct of research would not violate the principles of respect for persons, beneficence and justice," both conditions subject to approval of the National Ethical Advisory Board and "the secretary of the responsible federal department" (Recommendation 6). While the former provision can be satisfied on the grounds of actual scientific data, serious difficulties arise with respect to the latter, as the Commission does not engage in a more detailed ethical explanation of its understanding of the principles of respect for persons, beneficence, and justice, thus permitting a broad range of potential interpretations, which may not necessarily be fully protective of the incompetent subjects' welfare.

An explanatory statement attached to the Report and authored by half of the Commissioners, in which the topic of proxy consent, due to its intimate ties to informed consent in general, has been addressed under the principle of respect for persons, may shed some light on how to handle this issue. At the outset, the Commissioners observe that "the incapacity of children to consent is one aspect of a more general dependency on adults who are responsible for their care" (National Commission for the Protection of Human Subjects 1977b, 128), and maintain that "the permission that parents give for children's participation in research can be accepted as an exercise of their general role, as caretakers, to guide decisions affecting their children's lives and activities." At the same time, however, this role remains constricted due to the "developing autonomy of children." Consequently, dissent of the child should generally be honored unless "the research involves a therapeutic intervention that is unavailable outside the research context" (National Commission for the Protection of Human Subjects 1977b, 129). To support this assertion, the Commissioners refer to the tendencies observable among lawyers, philosophers, and pediatric researchers, all of whom consider the child who has reached the age of reason, usually said to be seven years, as sufficiently mature to offer binding assent or dissent for research. Yet this practice, as the Commissioners admit, even if useful, does not contribute toward arriving at a reliable judgment on the actual maturity of minor candidates being considered as experimental subjects. Moreover, it leaves unsettled the issue of enrolling in research children younger than seven years old. Needless to say, both issues become more significant in situations when, for example, parental judgment cannot be trusted as truly protective of the child's well-being.

9.3.1.1 The National Commission's Arguments Against Ramsey

Additional clues to the ethical and anthropological premises that underlie the Commission's recommendations can be uncovered examining chapter eight of the Report, which contains the Commission's critical review of the philosophical arguments offered in the course of the debate on proxy consent set off by Ramsey, May, and McCormick. The Commission criticizes the former's position, based on five objections.

First, at the heart of Ramsey's reasoning lies the conviction that, as already noted, minors must not be subjected to research precisely because of their incapacity for offering full-blown informed consent. While their participation can be justified in therapeutic settings, this certainly does not hold for the nontherapeutic milieu. To this the Commission responds that a significant difference, apparently overlooked by Ramsey, exists between (1) the refusal to participate in research on the part of competent individuals and (2) assuming such a refusal for children, perforce incompetent, and capable of neither meaningful consent nor refusal. While the dissent of minors should always be respected, it is also true that many children willingly assent to becoming subjects in research.

Second, although Ramsey's proposal is "internally consistent" in its unconditional ban on nontherapeutic research regardless of the risk at stake, it nonetheless "neglects discussion of the low level of risk involved in most research involving children" and also omits entirely the issue of children's assent. Consequently, in Ramsey's account, even research "that is only observational, or merely uses questions, or involves only paper and pencil tests or procedures of a routine medical examination" may not be permitted (National Commission for the Protection of Human Subjects 1977a, 96).

Third, the Commission believes that Ramsey builds upon the "false dichotomy between research intended to benefit subjects directly ... and research intended to develop more knowledge," while in fact "much research does not fit neatly into either category." The lack of precision in Ramsey's language unfolds also, as the Report sees it, in the way he employs such concepts as research and therapy.

Fourth, the Commission refers to the "thought experiment" Ramsey once made upon the research imperative, in relation to which he wrote: "if today we mean to give such weight to the *research imperative* ... then we should not seek to give a principled justification of what we are doing with children. It is better to leave the research imperative in incorrigible conflict with the principle that protects the individual human person from being used for research purposes without either his expressed or correctly construed consent." In such conflicts, Ramsey continued, "moral agents are under the necessity of doing wrong for the sake of the public good. Either way they do wrong. It is immoral not to do the research. It is also immoral to use children who cannot themselves consent and who ought not to be presumed to consent to research unrelated to their treatment. On this supposition research medicine, like politics, is a realm in which men have to 'sin bravely'" (Ramsey 1976, 21). In brief, the Commission questions Ramsey's support for "sinning on the side of avoiding harm rather than attempting to promote welfare," on the grounds that the harm-benefit ratio may actually favor a given research project rather than caution against it. Such, for instance, is the case of therapeutic research promising significant benefits at the cost of minimal risk. My problem with this objection, however, is that by arguing from the position of the therapeutic setting the Commission misreads Ramsey's point made, as the passage just quoted clearly shows, with respect to nontherapeutic research which by its very definition denies the existence of any direct benefits for the subject and thus precludes any risk-benefit calculus. Even if one extends the content of the term "benefit" so as to reach beyond a strictly medical meaning, a practice undoubtedly alien to Ramsey himself, a serious issue rapidly emerges regarding the incommensurability of risks and benefits to be evaluated and weighted.

Fifth, the Commission's Report touches upon the Kantian feature of Ramsey's thought. In a fashion somewhat resembling that of McCormick, the Commission concedes that the child serving in nontherapeutic research is treated as a means, but certainly not only as a means since "no investigator is free to use a child in any way he wishes." In fact, when only minimal risk is involved and yet considerable advantages for others are to ensue, it is "far from obvious that any immoral treatment is present" (National Commission for the Protection of Human Subjects 1977a, 99).

9.3.1.2 The National Commission's Arguments Against McCormick

After rejecting Ramsey's analysis the National Commission turns to McCormick's approach which, even though less restrictive and closer in conclusions to the Commission's own stance, nonetheless provokes its disfavor for the reasons outlined below.

First, McCormick's resorting to natural law proves problematic in the Commissioners' eyes because, to invoke their words, "natural law arguments have been subjected to sharp criticism in ethical theory." Although this is surely the case, it nonetheless does not follow, as the Commission seems to imply, that the natural law has thereby been eliminated from the realm of supportable ethical justifications. Nor does it follow that the arguments based on natural law are flawed. This of course is not to assert that McCormick's use of natural law is correct, but given the variety of natural law approaches, not even mentioned in the Report, in my opinion a more cautious stance could be expected of the Commissioners. In other words, what they actually attempt to criticize is not the natural law theory as such but rather the way in which it is used by McCormick, a distinction lacking in the Report. Moreover, the Commission's mention of sharp criticism of natural law is unsupported by any reference. The only appeal in the body of the text is made to "one common objection" in line with which "it does not follow from the wide or even universal sharing by human beings of certain values or purposes (e.g., health, happiness, etc.) that human beings ought to promote those values or purposes" (National Commission for the Protection of Human Subjects 1977a, 101).

Second, the very core of McCormick's argument as viewed in the Report becomes its Achilles' heel, as well. McCormick maintains, to invoke the Commission's wording, that "if one ought to do it [one thing or another], then consent may be validly assumed." In point of fact, however, "consent is precisely what may not be assumed even if one ought to do it. One reason why the requirement of informed consent has become so important in [then] recent years is that the consent of some subjects was never solicited, because a prior judgment had been made that they ought to participate" (National Commission for the Protection of Human Subjects 1977a, 103). Put briefly, the Commission contends that "there are probably numerous activities in which adults ought to participate but to which many would not consent" (National Commission for the Protection of Human Subjects 1977a, 101). If securing informed consent aims primarily at safeguarding the human subject's autonomy, it is obvious that different people will exercise divergent choices based on their personal evaluation of available options. This includes the right to refuse participation in research, even if others think that such participation ought not to be refused. "While we have a moral right to demand that individuals fulfill their obligations," the argument continues, "some obligations are created only by an individual's own commitments, and we often have no right to demand the commitment itself. Consent is such commitment, and absent the commitment no valid consent can be presumed" (National Commission for the Protection of Human Subjects 1977a, 102).

Third, the Commission observes that McCormick's reasoning ultimately leads to the elimination of the need for proxy consent. "If a child ought to do something and the obligation justifies the child's doing that thing," the argument runs, "then the consent of his parents could neither validate nor invalidate his participation." Thus, the permission given by parents or guardians becomes immaterial to the justification for employing minors as research subjects. In the Commission's own terms, "if the natural law justification is correct, it actually undermines the consent model by rendering it gratuitous" (National Commission for the Protection of Human Subjects 1977a, 104).

Fourth, the Commission further discusses the issue of the diminishing of the role of proxy consent in McCormick's position. In his later writings, McCormick designated the sociality of humans as the conceptual foundation supporting the existence of minimal duties toward others. In the case of children, the argument goes, because of their humanity, these duties, with the participation in nontherapeutic research among them, can be presumed. In the Commission's view, in this modified approach McCormick shifts from a "presumed consent" to a "presumed duty" position. Even if he allows for parents to be "empowered to consent for the child's participation whenever the child should be willing to be involved in the research," such consent cannot be taken as anything more than "merely a device to protect the child." Needless to say, when consent is thus construed, it plays a peripheral role in the ethical argument (National Commission for the Protection of Human Subjects 1977a, 104).

9.3.2 Stephen Toulmin and Victor Worsfold

The approaches advanced by Stephen Toulmin and Victor Worsfold may be characterized as variations, at least to a degree, of McCormick's proposal. While discussing the topic of fetal experimentation, Toulmin introduces "one small modification," as he calls it, to McCormick's position. Instead of seeking a reasonable presumption of what the child ought to consent to, if he could, one should rather ask what the child "could not reasonably object to," again, if he could (Toulmin 1976, 10–18). While a rather fine distinction, such a reformulation, Toulmin believes, eliminates the danger of imposing moral obligations upon minors, one of the most disputed parts of McCormick's work.

A related view is to be found in Worsfold, who substitutes Toulmin's negative wording by a positive one. As the Commission phrased it, "rather than holding that the child could not reasonably object, Worsfold's criterion is that the reasonable child would approve, in retrospect, an invitation to be involved in research" (National Commission for the Protection of Human Subjects 1977a, 106). Two additional elements play a decisive part in Worsfold's contention, both taken straight from John Rawls. First, the proxy decisions "are to be guided by the individual's own settled preferences and interests insofar as they are not irrational" (Rawls 1971, 249; quoted in Worsfold 1974, 154). Second, when knowledge of the preferences is unobtainable, the decisions must rest on "the account of primary goods" (Rawls

1971, 250; quoted in Worsfold 1974, 156), which according to Worsfold "suggests at the minimum that children should be consulted about their aims and preferences" (Worsfold 1974, 156).

The Commissioners consider Toulmin's and Worsfold's stances as two variants of "the reasonable standard theory," which they greet skeptically for several reasons, all revolving around the ambiguity of the term "reasonable". First of all, "reasonableness of nonparticipation" in research "can be judged only by reference to a person's reasons for nonparticipation." Since these reasons cannot be known in the case of children they "must be projected by others and a decision about the reasonableness of these reasons must be made by others." Consequently, any benchmark used for arriving at these judgments, as it does not come from the child himself, stands as "external to the standard of what the child would do if he could choose." Furthermore, there is also uncertainty as to what exactly counts as the standard of reasonableness in question. Should it build upon the relevance of the knowledge to be achieved, or perhaps upon the values cherished by the research community, or rather upon the personal convictions of parents of the potential subject (National Commission for the Protection of Human Subjects 1977a, 107)?

More important and apart from these uncertainties, the Commission points out that "the mere lack of a reasonable objection does not justify appropriating others." In other words, while an absence of reasonable objection on the part of a person proves to be the sine qua non for employing that person in research, it certainly should not be taken as a sufficient justification or as the sole factor involved. A research project, even if there are no reasonable objections to it, may still be subject to questions regarding its ethical soundness. Finally, the reasonable-standard approaches create similar difficulties to that of McCormick because they also are too broad. Therefore, "if lack of reasonable objection or reasonable presumption of a later agreement justifies appropriation, then it justifies drafting adults as well as children" (National Commission for the Protection of Human Subjects 1977a, 108). This in turn undermines the importance of proxy consent by rendering it a mere procedural device only peripherally involved in the justification of using minors in experimentation.

9.3.3 William G. Bartholome

Another attempt to settle the issue under review was made by William Bartholome. Bartholome takes up the conviction once expressed by Beecher according to which, by subjecting their children to nontherapeutic research bearing the promise of significant benefit for others, the parents may actually discharge the duty "to inculcate into their children attitudes of unselfish service" (Beecher 1970, 63). Bartholome submits that the justification of proxy consent in nontherapeutic research hinges upon the moral development of the child. With this approach, his position has the merit of avoiding Ramsey's objection (of Kantian provenance) that subjects are treated in such research as means only, because the child's welfare (moral growth) is clearly an end also (Bartholome 1977b, 17) .

It is noteworthy that Bartholome's premise supports to a degree Ramsey's contention that minors can be subjected to nontherapeutic research only when they benefit from it. The crux is, however, that Bartholome extends the concept of benefit so as to denote also a moral benefit in addition to a strictly medical one. In this view, the covenantal bonds existing between parents and their children include not only the responsibility for the protection of the children but also the responsibility for augmenting and directing their moral growth. This goal, Bartholome contends, can be achieved primarily by example and education, which require close cooperation on the part of the parents up to the point of becoming "joint-subjects" in research (Bartholome 1977b, 18). In contrast to McCormick, however, he stresses that allowing children to participate in nontherapeutic research, rather than being an obligation that ought to be fulfilled, constitutes only an option, one among others, that parents may choose as a means to the end of enhancing their children's moral development. Yet this option has its own limitations as well. Particularly in the case of children who have not yet reached the age of reason, participation in nontherapeutic experimentation yields no moral benefit at all (Bartholome 1977b, 17). Consequently, such participation proves ethically unsupportable, especially when the medical procedures to be carried out are highly invasive.[170]

Bartholome's account finds favor with the National Commission because it calls to mind that "psychological and moral benefits may be derived from participation in research." At the same time, however, it seems "subject to objection that whenever it could not reasonably be said that a child would be instructed, the research could not be justified" (National Commission for the Protection of Human Subjects 1977a, 112). Still, it is precisely at this point that Bartholome's interpretation approximates that of Ramsey. While for the latter, research with minors must be disallowed unless it contains immediate therapeutic benefit, the former rules out research unless moral benefit is involved. As a result, Bartholome's approach not only excludes from nontherapeutic research "uninstructable" minors such as infants, but it also fails to explain why their participation in such research cannot be justified on other grounds.

9.3.4 The Early H. Tristram Engelhardt and Robert M. Veatch

Veatch and early Engelhardt are the last pair of authors whose views on the matter drew the Commission's attention. Even if to a degree they differ from one another at the level of ethical justification, both Veatch and Engelhardt reason within a qualified utilitarian framework and both concur that enrollment of minor human subjects in nontherapeutic research can be justified by the benefits flowing from such research to the whole class of children.

Engelhardt begins by noting that Ramsey and McCormick advance their reasoning on the supposition that proxies need to take into consideration "the will and wishes of incompetents." The point of being incompetent, however, is that one is not capable of morally willing something or of expressing binding wishes. No meaningful choice, in the moral sense, can be exercised by the incompetent individual, and this necessitates resort to proxies. Consequently, "infants, though often willful, have no free will, and are not the object of respect in the sense that adults are."

Even though the principle of respect for persons, the tenor of which for Engelhardt boils down to safeguarding autonomy, does not apply to infants, nevertheless their physical and psychological integrity do require protection. In other words, "if proxy consent is not seen as a safeguard against the violation of an infant's wishes (on the grounds that it is not a free agent as normal adults are), but rather as a means to protect its best interest, it follows that that experimentation that does not involve an increase of risk over the ambiance or any significant discomfort, should be allowed" (Engelhardt 1979, 33). Such experimentation is "prima facie proper" also because of the obligations which lie with all members of society.

Unlike McCormick, Engelhardt argues that instead of imposing the ought language upon the child, one should rather ask what risks the members of society may reasonably be required to undertake in the pursuit of projects that do not contain serious burdens and that are reckoned as relevant by society. In response, Engelhardt notes "that nearly all societies recognize a certain minimal level of risk as a social obligation to which one may be said to have implicitly consented by one's presence." Therefore, when nontherapeutic research of an authentic social benefit involves "very minimal risk" and "minor discomfort" nothing stands in the way of approving it on the grounds of "the minimal duties that each of us owe to our society." By saying this, however, in defense of his declared opposition to imputing moral obligations to infants, Engelhardt nonetheless ends up imputing precisely these obligations. There is also one more difficulty inherent in his argument, namely that of how to determine, be it at the individual or at the institutional level, the ceiling for "the very minimal risk" just invoked. Engelhardt sees "no clear lines to be drawn" in this regard yet hopes "that such lines will be drawn by prudent persons" (Engelhardt 1979, 34).

Veatch begins by classifying small children as "the clearest example" of the impossibility of informed consent due to their inability to exercise self-determination. For this reason, in line with Ramsey, May, and the National Commission, he takes it as "a mistake to speak of 'proxy consent' for experiments in children" (Veatch, 39). It is rather the consent of parents offered on the grounds of their own self-determination which allows for a certain amount of discretion with respect to choosing their children's education and inculcating into them the parent's own values. Still, the parents' authority may be limited by the courts of law, when the values relayed to their offspring depart too radically from those accepted in society. Nevertheless, one clearly legitimate way of exerting parental authority expresses itself in "encouraging the child to make minor contributions to the general welfare for the welfare of specific others" (Veatch, 41).

This statement relates closely to Bartholme's moral-development position but Veatch moves a step further and asserts, in accord with McCormick, that given the child's membership in society, some minimal duties he or she owes to society may be presumed regardless of whether or not factual consent was procured, still provided that the moral status of the child as "an end in himself" and bearer of individual rights is respected. Veatch, like Engelhardt, takes the mere prospect of significant social benefits to be achieved in the course of nontherapeutic experimentation holding no more than minimal risk as an ample basis for justifying subjection of children to such research. In his own words, "there must be some contribution to the general welfare which can be expected without consent which the reasonable person would find required." Put another way, "it is reasonable to treat the individual, nonconsenting subject as a means to an end under very limited and circumscribed conditions" (Veatch, 42) although because of "great dangers" inherent in implementing this conclusion within the purview of research, the authorization of parents becomes compulsory.

In spite of suggesting this safeguard, however, Veatch specifies neither why treating the incompetent as a means is reasonable, nor whether such treatment implies reducing the subject to a means only. In addition, by incorporating the reasonable-person standard, Veatch exposes his position to the very same critique that the Commission offered in regard to Toulmin's and Worsfold's approaches discussed above. This observation becomes even more relevant when one realizes that the reasonable-person standard also plays an essential role in Veatch's approach to older children, whose assent and refusal, although neither free nor informed, should nonetheless always be respected. Why so? Because, as Veatch's answer goes, "since the child has nothing to gain, it seems *reasonable* that his uninformed refusal should nevertheless be determinative" (Veatch, 42, emphasis added).

The inclusiveness of Veatch's account is simultaneously advantageous and disadvantageous. No doubt, by integrating the views of other scholars into his own position, Veatch demonstrates a high level of familiarity with the subject matter, combined with an attempt to work out an optimal position. Yet on the other hand, it is precisely by sharing the views of others that Veatch exposes his analyses to the very same charges leveled against them.

Apart from this, both Engelhardt's and Veatch's positions invite the criticism of the National Commission because their approach justifies a relatively high level of research participation with possible detriment to the welfare of individual subjects, the objection already made regarding Toulmin and Worsfold. In particular, the absence of clear specifications regarding the scope of research considered legitimate in Engelhardt's and Veatch's accounts raises questions. As an illustration, the issue of minor increase over minimal risk may be considered, i.e., whether or not such an increase can be justified in nontherapeutic research and if so, under what conditions. It seems that both authors simply assume that minimal risk delineates the boundary for ethically supportable experiments with human subjects without engaging in a further quest for more basic rationales backing this assertion. What is more, even if aware of the tensions characterizing the relationship between protecting individual subjects and pursuing social benefits, Veatch and Engelhardt fail to strike a proper balance between these two obligations.

This issue exemplifies, according to the Commission, a common danger present in all qualified-social-benefit approaches to proxy consent in nontherapeutic research, that is, such approaches "may justify so much on the grounds of the principle of beneficence that the principle of respect for persons fails to be applied." Although Veatch and Engelhardt both seek to escape this trap, their positions demand more advanced conceptual work in order to provide clearer criteria for the stands they embrace. As the Commissioners put it, "it is one thing to argue that some research on infants may be allowed, and another to develop the precise conditions under which it is justified" (National Commission for the Protection of Human Subjects 1977a, 33).

Chapter 10
Analysis

This chapter will first take up a critical revision of some premises endorsed in the course of the debate outlined above. It will then turn to the most recent arguments on the matter and will seek to lay the foundation for a sound ethical and anthropological rationale, justifying the qualified permissibility of employing incompetent subjects in experimentation that, while holding out no immediate benefit to them, redounds to the good of others.

10.1 The Debate Revisited

In addition to the critical remarks made earlier with respect to the arguments advanced in the course of the debate on justifiability of proxy consent in nontherapeutic research, this section will highlight three other problematic areas permeating these accounts: the position of autonomy, the concept of reasonableness, and the scope of parental authority over children. All three issues affect the way in which proxy consent is perceived and applied in the realm of nontherapeutic experimentation.

10.1.1 Overrated Position of Autonomy

The first flaw comes down to the overrated position reserved for autonomy, best seen in the accounts of Engelhardt and Veatch. Even the National Commission appears more anxious to rest its understanding of the principle of respect for persons exclusively on autonomy as compared, for instance, to the Belmont Report, although the extent to which this apparent tendency corresponds to the Commissioners' actual intentions remains debatable. As discussed in the first chapter of this study, even though *Belmont* overtly associates the principle of respect for persons with autonomy, it nonetheless leaves room for other components to be included under this principle as well. Although not specified by the Commission these components could offset and complement autonomy and thus render the Commission's account more acceptable.

G. Mazur, O.P., *Informed Consent, Proxy Consent, and Catholic Bioethics*, Philosophy and Medicine 112, DOI 10.1007/978-94-007-2196-8_10,

Unfortunately, no satisfactory evidence in this regard can be found in *Research Involving Children.* In its review of Engelhard and Veatch, the Commission actually assumes and operates on a specious dichotomy between autonomy and beneficence, which, as demonstrated in a previous chapter, stems from a self-determinative understanding of autonomy disconnected from the objective good and attendant on a paternalistic rendering of beneficence. Thus autonomy and beneficence, devoid of their content, are in conflict with one another. Accordingly, the Commission comments favorably on the fact that Veatch and Engelhardt include "an explanation of the proper balance to be struck between the *competing obligations* to respect persons and to benefit those in need of help" (National Commission for the Protection of Human Subjects 1977a, 118, emphasis added). The presumption is clear: autonomy (constitutive of respect of persons) and beneficence are in competition and must be balanced against each other.

The dichotomy between autonomy and beneficence also informs the way in which assent and dissent of the child are understood. The parents or guardians aim at the welfare of the child, which is to say they act on the side of beneficence. But when the child reaches the age of reason and becomes capable of autonomous choices he can thereby balance the parental influence. The absence of a firm objective footing upon which both autonomy and beneficence could be based is characteristic of all the approaches examined in this section, and inevitably leads to the acceptance even of those decisions of the child that are arbitrary, emotional, or the result of mere whim or caprice. Safeguarding the child's real good never comes down to merely protecting self-determinative autonomy. If it were so defined, the scope and significance of parental responsibilities would be diminished. Furthermore, this would fail to account for the dependency of the child. This dependency calls for parental guidance, which in turn may necessitate overriding the child's choices, regardless of the degree of autonomy they contain. Notably, this holds true for both therapeutic and nontherapeutic research, a point largely overlooked by the Commissioners and by those participants in the debate whose views have been examined in this section.

The philosophical critique of autonomy with its anthropological and ethical implications offered in a previous chapter of this study applies well in the context of the debate under review. An approach to autonomy that is conceptually impoverished, content-thin, and intuitive makes it unreliable as a rationale on which to build informed consent in general, and its proxy form in particular, in both therapeutic and nontherapeutic settings. Similarly, the alleged split between autonomy and beneficence cripples proxy consent offered for nonautonomous persons, for example, infants and small children below the age of reason, whose autonomy, because nonexistent, cannot be protected. Consequently no balancing between autonomy and beneficence can take place, as the welfare or best interests of the child can be safeguarded only by recourse to the latter.

As already set forth, self-determinative autonomy tacitly permeates BIS, making it subjective overall. Therefore, even if it is said to safeguard the best interests of nonautonomous persons, informed consent can at best protect the right of the parents to decide autonomously on behalf of their nonautonomous offspring, an illustration of which can be found in Veatch. No meaningful protection of the child's welfare is possible because without the concept of objective good, no one knows in what this welfare actually consists. Obviously, one can restrict parental discretion to distinguish good from evil in regard to their children, and indeed such attempts are made by courts in situations where blatant violation of children' rights takes place. But how to make a moral judgment regarding less obvious infringements upon the rights of an incompetent research subject? How to educate and guide proxies and other parties involved in surrogate decision making so that they can serve the true good of the subjects instead of merely exercising their autonomous rights or intuitively surmising what counts as the subjects' best interest? Certainly, in order to avoid the trap of subjectivism and arbitrariness, one needs to anchor these efforts in the objective account of human good derivative of human nature and its inclinations. Such an account supplies proxy consent with content and thus provides reliable guidance for proxies themselves.

10.1.2 Reasonable-Person Standard

As noted above, this standard is found in the positions of Toulmin, Worsfold, and Veatch. With respect to the first two authors, the National Commission correctly judged their use of the standard too obscure to serve as the foundation on which to build the justification of proxy consent in nontherapeutic research. Yet, in its critique of Toulmin's concept of reasonable objection along with Worsfold's proposal of reasonable approval, the Commission itself, too, falls prey to subjectivism. At one point it states, for instance, that "the reasonableness of nonparticipation in an activity that is primarily charitable or for the benefit of others can be judged *only* by reference to a person's reasons for nonparticipation" (National Commission for the Protection of Human Subjects 1977a, 107, emphasis added). It is precisely the word "only" that puts in question the accuracy of the Commission's statement because it suggests that no external standards can be applied in the evaluation of the reasonableness of the person's decision. While the person's reasons doubtless are important and must not be dismissed rashly, it still remains true that objective criteria of what is good and bad shed precious light on a person's action. Otherwise one could never meaningfully evaluate other persons' choices. Much less could one assess what is reasonable to do for an infant incapable of choosing or for an incompetent adult incapable of offering reasons for his choices. With respect to this, the Commission proposes that "possible reasons for nonparticipation *must be projected* by others, and the decision about the reasonableness of these reasons must be made by others" (National Commission for the Protection of Human Subjects 1977a, 107).

At first glance it seems that there is not much to question in this formulation, but upon closer inspection the following issues arise. First of all, the Commission does not specify on what grounds the reasonableness of the possible reasons for nonparticipation (and, by extension, also participation) in nontherapeutic research are to be judged by others. No mention is made of the child's good or of any other objective measures. Second, it is not clear what the word "projected" used by the Commission actually means. In my view, it approaches much too closely one of the arguments of McCormick's proposal, because to say that the reasons "must be projected by others" seems similar to saying that they must be "constructed off from what others ... *ought* as humans to do" (McCormick 1975, 128). Interestingly, the "ought" language applied by McCormick, although questioned by the Commission and removed from prominence in its assessment seems nonetheless to be implicit in its argument, at least as far as the proxies' duties are concerned. In the Commission's terms, the reason for participation *must* be projected, which I would take to mean that the proxy has an obligation to project it.

Obviously, there is nothing wrong in admitting that proxies have moral obligations toward their wards. But in what do such obligations consist? Is it only a matter of the proxy's autonomous right to do what he or she thinks is reasonable? Does this right justify expanding the legitimacy of proxy consent not only to minimal-risk research but also to research involving a slight increase over minimal risk, as the Commission does? Regrettably, the answers to these questions are not included in the Report. No doubt, they would be easier to ascertain if the Commission elaborated at greater length on the meaning and scope of the projections it proposes. Without such clarification, one can plausibly argue that the Commission, even though careful not to impute moral obligations to children, nonetheless implies some sort of imposition of an adult world upon children, the very argument characteristic of McCormick's position. Based on the Commission's analysis it is hard to see how otherwise to explain the reference to the reasons that *must be projected* by adult proxies and offered as a part of their consent or refusal for subjecting children to nontherapeutic experimentation. What is more, since the only basis for such a projection suggested in the Report comes down to a subjective grasp of reasonableness, detached from any meaningful concept of good, the Commission does not seem to do much better than Toulmin and Worsfold, with whose views it finds fault. These views reflect Veatch's understanding of the term "reasonable" and therefore, along with the Commission's argument, become subject to the critique of the reasonable-person standard carried out in the previous chapter of this study.

10.1.3 The Scope of Parental Authority over Children

This analysis centers largely on Ramsey's understanding of parental authority and its consequences. It also refers critically to other views, including that of McCormick. No doubt, the unqualified ban on subordinating children to nontherapeutic research as advanced by Ramsey exhibits a great deal of sensitivity and caution against potential abuses to which such subordination may lead. Therefore, it deserves recognition and cannot be rashly dismissed. On the other hand, however, it can be strongly challenged as too stringent and thus overly restricting parental guardianship over children. No fewer than three lines of reasoning support this objection.

First, excluding nontherapeutic research with children from the array of morally acceptable options inevitably results in denying children access to the possible beneficial results of such research. Although aware of this problem, Ramsey nonetheless did not seriously address it and suggested only that a remedy to leaving children as "therapeutic orphans"[171] is sinning bravely. As noted above, at the root of this recommendation lies Ramsey's conviction that in certain situations "moral agents are under the necessity of doing wrong for the sake of the public good." Put differently, "it is immoral not to do the research" but on the other hand "it is also immoral to use children who cannot themselves consent and who ought not to be presumed to consent to research unrelated to their treatment" (Ramsey 1976, 21).

Although the Commission's rebuttal of this conclusion seems to miss the point, Ramsey's inference can be questioned on other grounds. Above all, it is striking that Ramsey's own premises lead him to a situation with no clear avenue of escape. The ambiguity and flexibility with which Ramsey pursues his project, a factor addressed earlier in this study, are not without impact here. This weakness also encompasses the lack of clear distinctions between therapeutic and nontherapeutic research as noted by the National Commission. One may suppose that to minimize the damage caused by the strictures on nontherapeutic research Ramsey allowed for the epidemics exception and also extended the scope of therapeutic research so as to embrace the possibility of remote benefits for children. Combined with the absence of a clear notion of objective good to underpin his analyses, all these issues call into questions the accuracy of Ramsey's conclusions. Indeed, the suggestion that one is to sin bravely and that one in certain situations must do wrong is just as surprising as it is unacceptable. As such it not only leaves the dilemma in question unresolved but indeed it also suggests that no resolution is in fact possible. Thus it offers very little guidance for proxies. Finally, since it gives no clues on how to determine borderline situations that necessitate doing wrong, it is likely to result in confusion and subjectivism.

The second line of critique of Ramsey's view of parental authority develops along the lines of the notion of minimal risk. I believe that McCormick was right in pointing out that determining whether or not the amount of risk to be present in a given research project falls within the range of minimal is chiefly the task of the scientific community, an assertion to which Ramsey also subscribed but in a rather noncommittal manner. McCormick was also right in taking the lack of familiarity with what level of risk actually stands as minimal to be likely to generate exaggerated concern. Although Ramsey for the most part discounts this charge because "offensive touching," as he understands it, can come about regardless of the risk involved, he nonetheless betrays a fairly modest grasp of the possible threats proper to minimal risk itself.

Perhaps the best way to illustrate this observation is to recall the "accordion" objection he levels against McCormick. The image of the accordion is surely exaggerated, as the gap between the hypothetical zero-risk situation and one that features only minimal or hardly discernible risk is too narrow to allow for the amount of expandability and flexibility Ramsey suggests. As Friedman-Ross puts it, minimal-risk research "presents no more risk than that which is encountered in many activities to which parents typically expose their children for educational purposes. Many activities in a typical child's life present greater risks, including such routine activities as the participation in contact sports and travelling in the family car. Not only is it impossible to live in a risk-free world, but also it is contrary to the pursuit of a meaningful life plan" (Friedman Ross 1998, 92). This common-sense observation is closely related to the argument put forward by Grisez and explored earlier. Its conclusion is that although parents regularly expose children to minimal risk in situations not related directly to the child's welfare, no one questions their parental authority. Thus, no evident reasons exist why this authority is to be questioned in the research setting. Grisez would certainly not agree with Friedman Ross that being driven in the family automobile constitutes more than minimal risk for the child. By way of reminder, in Grisez's account, minimal risk denotes any risk that does not extend beyond life's common risks. Thus, driving a child in a car to do shopping for the grandparents, for instance, stays at the level of minimal risk.

I subscribe to this explanation and regard Friedman Ross's attempted distinction between minimal and greater than minimal risk to be arbitrary, as implied in the illustration just invoked. Indeed, she does not clarify why travelling in a family car is to be considered more risky than other routine activities in which children participate with their parents' permission. In contrast, active engagement in contact sports, another example offered by Friedman Ross, is more likely to exceed minimal risk, but since it directly brings about a number of benefits to the child, one cannot compare it to nontherapeutic research. Consequently the contact-sports example does not support the justifiability of proxy consent for nontherapeutic research featuring a slight increase over minimal risk, a conclusion one could draw from the way Friedman Ross puts her case. Such justifiability has been endorsed by the National Commission and will be revisited in the final chapter of this study.

Although more realistic than Ramsey in his stand on minimal risk, McCormick nonetheless fails to work out an appropriate explanation of parental authority. It is of interest that he moves from the position of parental authority to that of parental responsibility which, in turn, confines itself to ascertaining and honoring what the child ought to do. From this perspective, it is the child who "*ought* to want" (McCormick 1981b, 64) participation in research insofar as it poses no threat to his good. In such cases parents ought to accept the "want" of the child. It is not clear, however, whether or not parents should want it, too. In fact, McCormick's approach implies that what parents want actually does not count for much because it is possible to find out the child's duty without parental assistance. Thus the authority of parents becomes diminished, if not essentially relegated to an insignificant secondary factor in proxy decision making.

The third dubious feature of Ramsey's approach to parental authority concerns the reduction of the potential benefits to be gained from participation in nontherapeutic research to those that are medical only. To some extent this reduction is suggested by the very distinction between therapeutic and nontherapeutic research. After all, the word "therapeutic" implies that the patient is helped *medically*. By contrast, the term "nontherapeutic" precludes such help. Yet, it is not evident, to say the least, why the latter has to prevent a participant from gaining non-medical benefits as well. Even without a therapeutic component, research can still yield benefits otherwise valuable to the subject.

It is here that Ramsey's account falls short. At one point, while arguing contra McCormick, Ramsey states that the only good to which childhood inclines is "the preservation of life and a healthy life and growth" (Ramsey 1976, 22). Does this imply that such goods as moral development do not count? Put another way, why must healthy life and growth confine itself to merely physical or medical aspects? Certainly, moral growth of the child does not constitute a lesser good, to be pursued under parental guidance. Even if not entirely ignored by Ramsey, this good was nonetheless treated only peripherally in his work.

An illustration of this is his discussion of the debate on organ donation in *Patient as Person.* While examining the positions of Martin Nolan and J. E. Murray, according to which organ donation meets standards of moral soundness on the strength of the principle of totality, Ramsey finds it insufficient to say that even if a donor suffers physically as a result of her donation, her act is nonetheless justified on the grounds that she benefits from such a donation in a spiritual way. Ramsey disapproves of such use of the principle of totality as "strange" and labels the reasoning it features as a "sticky benefits theory" (Ramsey 2002, 181). Consistent with his stand on research, he also recommends "free and informed consent on the part of the donor" as a better foundation on which to justify organ donation (Ramsey 2002, 187). This line of Ramsey's thought must have been unknown to the National Commission when it faulted him for overlooking the existence of other-than-medical benefits possibly involved in nontherapeutic research (see National Commission for the Protection of Human Subjects 1977a, 112). Clearly conscious of such benefits, Ramsey nonetheless regards them to be indecisive for justification of nontherapeutic research and organ donation. In effect, he not only prevents the whole class of children from access to possible future benefits of nontherapeutic research but also fails to consider the developmental needs of minors, including learning altruism, maturing in virtue, and forming conscience. Likewise, the social character of human nature in which children share has been dismissed altogether.

Even though McCormick's interpretation of this, as convincingly argued by May, is unsound, it does not follow that sociality constitutes an irrelevant aspect of childhood. Interestingly both Bartholome and Veatch are more appreciative of the significance of both moral development and the sociality of minors than is Ramsey. Veatch, nonetheless, goes too far when he not only (and rightly) sees the promotion of moral growth of the child as properly appertaining to parental authority but also emphasizes the need for encouraging children to contribute to the welfare of others. As May explained, however, precisely because they are not moral agents, children cannot pursue the common good. I believe that one should clarify this basically correct observation by adding that in the case of older children, including so called emancipated minors, contribution to the common good is viable.

10.2 Toward a Sound Solution

The analyses contained in preceding sections reveal a number of inadequacies in the way the authors studied account for the justifiability of proxy consent in nontherapeutic research. The goal of this section is to explore more recent approaches that set the stage for the resolution of the issue, to be provided in the next and final chapter. I will proceed by presenting the views of the later William. E. May, then Benedict Ashley and Kevin O'Rourke, and finally the teaching of the Magisterium.

10.2.1 The Later William E. May

May's long-held position, firmly retained for 30 years, ruling out as morally unsupportable all types of nontherapeutic research with minor subjects, changed in 2002. He was, as he put it, "sharply challenged" by the participants of a colloquium that took place that year under the auspices of the Pontifical Academy of Life with the aim of setting the stage for the Academy's plenary session to be held early the following year. It was at this meeting that May first came to realize that "although McCormick's *argument* to justify proxy consent on behalf of 'voiceless' subjects in the nontherapeutic situation may be severely criticized for the reasons given, the position he took can be defended on other grounds." Indeed, "it would not be unreasonable—and therefore not contrary to objective moral standards—if parents, for instance, were to allow experimentation involving their children for the benefit of others if the experiment involves no significant risk" (May 2000, 224). In accord with Grisez's depiction of minimal risk as including the level of risk not exceeding that present in common daily-life activities, May submits that no convincing reasons exist for this kind of risk to invalidate the children's participation in nontherapeutic research.

Still, the minimal-risk condition cannot be taken as the sole justifying factor for such participation. As May points out, three more requirements must be met: (1) serving as subjects may involve at maximum only slight inconvenience or encumbrance; (2) the research project must be highly favorable in the benefits to come and (3) it cannot be performed on populations other than the incompetent. By honestly satisfying these conditions parents avoid treating the child as a mere object of use which is a means only and thus they cannot be deemed derelict in their duty to safeguard the child's welfare (May 2000, 225–26).

Three additional traits of May's presentation deserve further attention. The first is his use of the word "reasonable" in a way that implies a different meaning from that based on self-determinative autonomy and criticized in the previous chapter of this study. Rather than referring to a vague term which is subjective, intuitive, and ultimately arbitrary in the sense of presupposing a predominant role of the faculty of will as artificially detached from both its object, i.e., good, and from the faculty of reason that knows the good and presents it to the will, May points to the objective moral order that is embedded in human nature and that safeguards true human good. Notably, the wording May uses is precise as he never explicitly states that it would be good or reasonable for parents or legal guardians to subject the children under their custody to minimal-risk nontherapeutic research. Rather, he argues that such an act "would not be unreasonable," that is, as I understand it, not contrary to the child's good, provided that no discernible risk is present and that the three requirements outlined above are fulfilled. In my opinion, caution in the choice of words here is relevant because some nontherapeutic experiments may actually neither directly foster the child's good nor violate it. The case of infants subjected to merely observatory procedures that hold out no direct benefit for them provides but one illustration in this regard.

Second, May holds a broader position in regard to decisions made by minors who have attained the age of reason. Although many such minors "can make free and informed decisions," and thus should be provided with adequate information about the experimental procedures, May nevertheless holds such decisions as subject to being overridden by parents when necessary in both therapeutic and nontherapeutic settings. Thus, in contrast to the National Commission's stand exemplifying a major trend in today's bioethics, according to which the assent and dissent of the child should always have binding force in nontherapeutic research, May assumes a broader position which, even though not expanded upon at length, proves consistent with the objective moral standards undergirding proxy decision making.

I believe that a plausible interpretation of May's approach can be advanced along the following lines. Despite the fact that the child, when old enough, usually arrives at a basic understanding of the nontherapeutic procedure, including its purpose, inherent risks, and inconveniences, she or he may still not possess a clear comprehension of his or her own objective good, and thus not be able to judge whether or not participation in this research counts as morally good (in harmony with that personal objective good), bad (contrary to that objective good) or neutral. Certainly, parents are more likely to exercise better insights in this respect, especially when other than medical benefits are at stake, by reason of which they should be able to override their children's preferences. Thus, May's understanding of the assent and dissent arises from the recognition of the real objective good of the child, the promotion of which appertains to the parents' immutable prerogatives. As such it stands in stark contrast to the autonomy-based models, which is disconnected from the human good and therefore forced to honor the child's wishes regarding participation in nontherapeutic experiments, even if these wishes have little or no objective foundation in fact.

Third, unlike McCormick, May refrains from making assumptions about the child's moral obligations. Yet, on the grounds of the principle of fairness[172] he now allows for such assumptions as far as competent research subjects are concerned. In other words, after changing his position, May came to terms with McCormick in that consent can be morally required in nontherapeutic research featuring no more than minimal risk, provided that the inconveniences it involves are negligible. As May phrases it, "in an instance of this kind, one can easily do something of benefit to fellow human persons with no cost or minimal cost to oneself." Therefore, "it seems that in fairness one could have a *moral obligation* to help out" (May 2000, 222). This does not entail, of course, that such obligation can be imputed to small children (below the age of reason) because as they are not yet moral agents they simply cannot be obliged to act morally (May 2000, 223).

10.2.2 Benedict Ashley and Kevin O'Rourke

In the fourth edition of their *Health Care Ethics*, Ashley and O'Rourke refer to the debate on justifiability of proxy consent in nontherapeutic experimentation and embark upon their own explanation. They note that proxy decisions "must be made in view of the good of the individual person, not for a higher good, for a class good, or for the good of another person, which would amount to manipulation of the person as a mere means" (Ashley and O'Rourke 1997, 350). This is the first principle to be observed in research. The second, which undergirds the responsibility that proxies have for incompetent subjects, is that of human dignity in community which "affirms that no person can achieve fulfillment without sharing in the common good and contributing to it" (Ashley and O'Rourke 1997, 351). As noted previously, although in their understanding of the principle of human dignity in community, Ashley and O'Rourke follow Aquinas' account of the four goods to which all human beings find themselves naturally inclined (the fourth good is living in society), they nonetheless do not demonstrate satisfactorily how this principle works as applied to proxy decision making. In addition, by stating that respecting human dignity expresses itself in "the maximal, integrated satisfaction of the innate and cultural needs, biological, psychological, ethical, and spiritual, of all human persons, as individuals and as members of both their national communities and the world community," Ashley and O'Rourke introduce a certain degree of ambiguity. The reason for this is twofold.

First, further clarifications of the statement cited are necessary in view of the fact that all human needs can never be satisfied equally, much less maximally. It is often necessary to sacrifice some of them, especially lesser ones, in order to satisfy others, in particular those of a higher order. Moreover, because what is referred to as needs in ethical discourse may imply purely subjective content, and hence may not point to needs in the sense of an absence of or a desire for something objectively good that calls for satiety, additional clarifications are needed. But even if an objective meaning is retained, a hierarchy of goods and corresponding needs still has to be recognized. As is case with the needs, numerous goods also have to be sacrificed so that the pursuit of other goods might be possible.

Second, Ashley's and O'Rourke's reference to human needs, although useful in therapeutic research, offers little help in nontherapeutic settings with the incompetent. By its very definition this research does not respond to the needs of the subjects involved but rather to the needs of others. This, of course, may not be entirely true when some degree of capacity to consent exists on the part of the subject, as often is the case of minors who reached the age of reason and are capable of gaining advantages other than medical ones from participation. Still, a more comprehensive account of objective needs, rooted in human nature, particularly the need to live in society which is essential to the principle in question, proves indispensible for making this position fully workable.

In spite of these imperfections, however, both the principle of the individual good of the research subject and the principle of human dignity in community as laid out by Ashley and O'Rourke provide a sound basis for justification of proxy consent in nontherapeutic research with no more than minimal risk entailed. They stress that in such settings the guardian cannot act on the basis of "the presumed consent of the ward, which is merely hypothetical, but rather must do so on the grounds of the actual need of the ward for care." Therefore, the argument goes, "theories of presumed consent based on what the ward *ought* to do if the ward *could* consent are weak." This conclusion notwithstanding, our authors consider participation in minimal-risk nontherapeutic research as permissible for incompetent individuals on the strength of the adage of classic moral theology in line with which "little counts for nothing." By preferring negative wording, like May, they use caution while introducing this qualified permission. As they put it, by subjecting the incompetent to minimal-risk nontherapeutic experimentation "the guardian does not fail in his or her responsibility to care for the ward" (Ashley and O'Rourke 1997, 351).

Even though Ashley and O'Rourke do not enlarge upon the scholastic maxim to which they refer, a possible method of interpreting it can begin, in my estimation, by pointing out that minimal risk does not in itself oppose the good of the incompetent subject, precisely because it is minimal or little, which is to say, negligible and trivial. In fact, it is by seriously taking the subject's good into account that the guardians come to realize that although at times it may not be enhanced by participating in a nontherapeutic experiment, this good will not be endangered, either, unless more than minimal risk comes into play. Such an interpretation is in accordance with Ashley's and O'Rourke's first principle according to which proxies must always act "in view of the good of the individual person."

An additional, and highly significant, line of reasoning justifying proxy consent in nontherapeutic research, put forth by Ashley, draws from the sociality of all human beings. In brief, all persons, including infants, "constantly profit from sharing with others. If a child is benefited by clean air which it has done nothing to improve why can it not, without loss to its human dignity, be used for some social purpose? It will profit itself from the fact that it is part of a healthy society improved by research."[173] By posing the "why-not" question to which, indeed, no plausible negative answer seems to exist, Ashley attempts to demonstrate indirectly the validity of proxy consent in nontherapeutic settings with no more than minimal risk involved. I believe the merit of this approach can be described as twofold. First, by pointing to other-than-medical benefits, inherent in virtually all scientifically sound research, this approach avoids reducing the incompetent subjects to being means only and thus disarms one of the prominent objections issued by Ramsey. Second, by abstaining from recourse to the "ought" language employed by McCormick, Ashley's statement does not impose the adult world of social obligations upon minor research subjects.

Although they do not incorporate this social argument into the latest (fifth) edition of their *Health Care Ethics*, Ashley and O'Rourke reinforce the justification of subjecting the incompetent to nontherapeutic research somewhat tentatively suggested in the fourth edition, by simply declaring that "it is not licit to expose a ward to other than minimal risk in nontherapeutic research" (Ashley, DeBlois, and O'Rourke 2006, 117). It follows that as long as merely minimal risk is at stake such exposure may be permitted.

10.2.3 The Teaching of the Magisterium

In order to comprehend the Magisterium teaching on proxy consent to nontherapeutic research, one must (1) note a conceptual difference between research and experimentation operational in this teaching and then (2) examine such consent in regard to human subjects (a) not yet born and (b) already born.

Regarding the first issue, the explanation of the distinction between research and experimentation is to be found in the *Charter for Health Care Workers* (n. 82) and in the Instruction *Donum Vitae*. The latter document, which offers more detail on this matter, specifies that while research indicates "any inductive-deductive process that aims at promoting the systematic observation of a given phenomenon in the human field or at verifying a hypothesis arising from previous observations," experimentation denotes "any research in which the human being (in the various stages of his existence: embryo, fetus, child, or adult) represents the object through which or upon which one intends to verify the effect, at present unknown or not sufficiently known, of a given treatment" (part I, n. 4). In other words, as used in *Donum Vitae,* research encompasses observational procedures, whereas experimentation presupposes some kind of active intervention or manipulation on the human subject.

With this distinction in mind, *Donum Vitae* rejects all nontherapeutic experimentation on unborn humans by emphasizing that "*if the embryos are living, whether viable or not, they must be respected just like any other human person; experimentation on embryos which is not directly therapeutic is illicit*" (emphasis in text). Regarding research, however, the position taken by the Instruction is more complex. First, no explicit distinction between therapeutic and nontherapeutic research is made in the text of the document. Second, the Instruction allows for research on the unborn unless, "even if limited to the simple observation," it nonetheless involves "risk to the embryo's physical integrity or life by reason of the methods used or the effects induced" (part I, n. 4).

Still, it remains unspecified what amount of risk the Instruction implies here. Both the context of its assertion and the way it is phrased suggest, however, that risk to the subject's life and integrity means any threat or danger to the subject's life and integrity. Therefore, as I understand it, if research (i.e., a merely observational procedure) endangers neither the unborn subjects' life nor their integrity, it may be permitted. Yet this conclusion certainly does not hold for nontherapeutic experimentation with embryos and fetuses, of which *Donum Vitae* disapproves unconditionally. The reason underlying this disapproval, offered in the Instruction, points to the fact that "experimentation on embryos and fetuses always involves risk, and indeed in most cases it involves the certain expectation of harm to their physical integrity or even their death" (part I, n. 4). Again, this statement makes sense only when the risk to which it refers poses a threat to the subject's life or integrity. In my estimation, with respect to the unborn employed as subjects, it is not unreasonable to concede that even minimal risk can present such threat.

Three other Magisterium documents, the first two of which are referenced in both *Donum Vitae* and the *Charter for Health Care Workers*, unanimously and firmly share *Donum Vitae*'s disapproval. The first document is the 1982 address of Pope John Paul II to the members of the Pontifical Academy of Sciences in which the following words were contained: "I condemn in a most explicit and formal way experimental manipulation of the human embryo, because it is a human being; from the moment of its conception until death it can never be instrumentalized for any reason whatsoever" (John Paul II 1982a). The second document, which asserts that "respect for the human being excludes all kinds of experimental manipulation or exploitation of the embryo" (Holy See 1983), is the *Charter of the Rights of the Family*, published in 1983. Finally, the *Ethical and Religious Directives*, previously discussed, provide that "nontherapeutic experiments on a living embryo or fetus are not permitted, even with the consent of the parents" (n. 51).

In regard to proxy consent to nontherapeutic experimentation with human subjects already born, the *Ethical and Religious Directives* stipulate that "the surrogate can give this consent only if the experiment entails no significant risk to the person's well-being" (n. 31). In the same vein, the Code of Ethical Standards for Catholic Health and Aged Services issued by the Australian Bishops stipulates that "nontherapeutic experimentation must involve no significant risk at all" (n. 6.6). Does the gauge of "no significant risk" incorporated in both statements amount to "no discernible risk," that is, to minimal risk? Because this question is not answered in the negative in any of the Magisterial documents, I believe that the answer to this question is positive.

The *Charter for Health Care Workers* seems to support such an answer, when, after admitting that a total exclusion of risks from experimentation is not only impossible but indeed unwelcome because it would "paralyze all scientific research," it immediately adds that nonetheless, "there is a threshold beyond which the risk becomes humanly unacceptable. This threshold is indicated by the inviolable good of the person" (n. 78). As demonstrated above in the examination of the views of May, Ashley, and O'Rourke, the presence of merely minimal risk in research does not violate the subject's good, objectively understood. Also, in my view, the reason that *Donum Vitae* accords special attention to the distinction between solely observational research procedures and experiments that imply intervention or manipulation might consist precisely in that the former does not necessarily compromise the subject's good (life, physical integrity), even the good of such vulnerable subjects as embryos and fetuses, whereas the latter must be seen as very likely to do so. Consequently, one can deduce that the minimal-risk standard proves compatible with the Magisterial teaching on moral justifiability of proxy consent in nontherapeutic research with the subjects already born.

Commenting on this teaching, May observes that "one might reasonably question why all non-therapeutic experiments on the unborn are illicit because of the serious risks they pose and why such experiments are justifiable on human persons already born." Put another way, why "an unborn child cannot be a legitimate subject of a non-therapeutic experiment, but apparently the same child, minutes after birth, can be?" (May 2000, 228). In response, May highlights a great vulnerability of newly born infants by reason of which their status as potential candidates for subjects in nontherapeutic experimentation does not seem to diverge from that of fetuses. In other words, it is only when subjecting a newborn baby to nontherapeutic procedures that surely entail no significant risk that such actions may be approved. I would conclude that whether or not such situations occur in real life is a completely different issue and indeed one far beyond the scope of this study.

Chapter 11
Resolution of Issues Related to Proxy Decision Making

The content of this brief final chapter has been organized around two goals. First, the critique of mainstream accounts of proxy decision making and the minimal-risk controversy, both previously discussed, will be revisited in the light of the fundamental principles of objective morality as laid out by the Magisterium of the Catholic Church. These principles relate to the objective human good and aim at the protection of human dignity. As such they include objective truth, in particular absolute moral norms prohibiting intrinsically evil acts, together with conscience and virtue. While enlarging upon these principles, I will also refer to a number of major consent-related issues that impair both the human-rights documents and some of the analyses of informed consent addressed in previous chapters. In doing so, I will also provide a brief summary highlighting major moral issues detected in the course of this study.

My second goal is to propose a resolution to the problem of justifiability of proxy consent in nontherapeutic research. This proposal emphasizes the risk factors as well as the close relationship between the individual good of the research subject and the common good.

11.1 Moral Truth, Conscience, and Virtue

As noted earlier in this study, Pellegrino and Thomasma identify achieving truth as the goal of scientific research and then go on to say that "the good internal to practice of research is truth" (Pellegrino and Thomasma 1993, 135). Indeed, an intimate connection between the truth and the good must be recognized in research, as in any other human activity. As Pope Benedict XVI put the matter, "truth means more than knowledge: knowing the truth leads us to discover the good" (Benedict XVI 2008a). Therefore, *truly* scientific research is always "oriented to the good of the human being" (John Paul II 2004, n. 4) and the same may be said of free and informed consent, including its proxy form. In consequence, both proxies and researchers must never act so as to compromise or disregard the true good of the subject. Note, however, that neither the human-rights documents nor the accounts of proxy decision making critically examined in earlier chapters of this study meaningfully refer to

G. Mazur, O.P., *Informed Consent, Proxy Consent, and Catholic Bioethics*, Philosophy and Medicine 112, DOI 10.1007/978-94-007-2196-8_11,

the truth as a superior value of scientific research, much less to the objective good of research subjects as calling for respect and protection.

Properly understood, truth in research essentially encompasses two components. The first is the moral truth about humans expressed in the precepts of natural law, while the second points to the scientific truth that every soundly designed research project presupposes and aims to attain. These two dimensions of truth, moral and scientific, should coexist harmoniously in every research project. If scientific truth seems to clash with its moral counterpart, there is inevitably some level of falsehood involved. In relation to this, the Magisterium teaches that authentic scientific research never resists moral laws or faith because both faith and reason are derived from God.[174] *Gaudium et Spes*, for instance, underscores that "since the same God who reveals mysteries and infuses faith has bestowed the light of reason on the human mind, God cannot deny himself, nor can truth contradict truth" (n. 36).[175] Or, as Pope Benedict XVI put it, "the one who seeks the truth becomes the one who lives by faith" (Benedict XVI 2008a). Accordingly, "The Catholic Principles and Guidelines for Clinical Research" published recently in the U.S. (hereafter referred to as "the Principles and Guidelines")[176] debunk the stereotype that the Church impedes the advancement of scientific knowledge. As a matter of fact, "the Church does not create any barriers to truth" because genuine scientific truth never contradicts faith (preamble). What does oppose faith and impugn human dignity are wrongful methods of attaining knowledge. Yet this does not change the fact that "to seek the truth and to adhere to it, once it is known," constitutes "a prior moral obligation" (John Paul II 1993, n. 34). Still, "man does not create truth; rather truth discloses itself to man when he perseveringly seeks it" (John Paul II 1997). It follows that in surrogate decision making the truth is to be sought, not projected or fabricated, at both the scientific and the moral level, by both researcher and proxy.

According to the Principles and Guidelines, the principle of truth requires that "informed consent must be complete, clear, and communicated at a level that can be understood by the subjects in the research" (Guideline 1d). The communication constituent, as already discussed in previous chapters, encompasses counseling, which in turn relates to formation of conscience. Indeed, conscience and truth stand in very close relationship to one another. To invoke the teaching of Pope John Paul II, the dignity of conscience as well as "the authority of its voice and judgments derive from the *truth* about moral good and evil, which it is called to listen to and to express" (John Paul II 1993, n. 60). By definition, "conscience is a judgment of reason whereby the human person recognizes the moral quality of a concrete act" (*Catechism of the Catholic Church* 1997, n. 1778). The *Catechism of the Catholic Church* describes the human conscience as containing three levels. First is the level of the fundamental principles of morality (synderesis). The second level pertains to the application of these principles "in the given circumstances by practical discernment of reasons and goods." Finally, at the third level, one makes a "judgment about concrete acts yet to be performed or already performed" (n. 1780).

The distinction between "reasons and goods" characteristic of the second level requires the virtue of prudence. Is prudence then different from conscience? While certain thinkers, such as Joseph Pieper, answer this question in the negative and essentially propose to equate conscience with prudence (see Pieper 1966), others, as, for example, McInerny, take a different stand. According to McInerny, in contrast with virtue, "conscience is said to be an act, not a habit" (McInerny 1997, 105). Furthermore, while conscience operates on a purely cognitive level (one arrives at the knowledge of what good ought to be done), the moral agent still needs to relate to the cognitively grasped good personally, i.e., as the agent's own good to be sought and pursued. After all, one can be rightly advised by one's conscience as to what is morally good under a given set of circumstances but still choose to act otherwise, that is, contrary to the moral ideal (the good cognitively grasped). It is in practically realizing the moral ideal that one needs virtue. As McInerny puts it, "prudence or practical wisdom determines how the moral ideal can be realized here and now; that is, thanks to prudence we deliberate, judge, and command as to the means of realizing the end" (McInerny 1997, 107–108). In referring to McInerny, Anthony Fisher offers a helpful interpretation of the point in question, according to which prudence counts rather as "adjective (a quality of the virtuous mind)" or even more as "adverb (a quality of virtuous doing), while conscience is really a noun (the mind thinking practically) and especially . . . a *verb* (doing practical reasoning toward good action)" (see Fisher 2007, 5). Still, the mere performance of the action hinges less on the practical reasoning that precedes it and leads to it than on the moral agent's habitual dispositions.

McInerny's and Fisher's insights on the difference between conscience and prudence not only provide a valuable commentary on Aquinas' account of these matters but also serve well the purpose of this study. Certainly, just as it is for every individual, it is of the highest relevance for an investigator, research subject, proxy, and others involved in the informed-consent process to form their consciences well, i.e., in accord with the objective moral truth, or as the *Catechism* puts it, "in conformity with the true good willed by the wisdom of the Creator" (n. 1783). This conformity with the true good provides the best protection for research subjects, especially those without competence or with compromised competence. Moreover, it enables the proxies truly to care for their wards and helps the investigators to keep from introducing into the research project any elements of abuse or exploitation in the name of a superior interest contrary to the good of subjects, be it the interests of society, of science, or of any other entity. Because of its agreement with the objective good, the well-formed conscience never deteriorates to the level of private caprice, whims, or taste. Nor can it be "reduced to social advantage, to group consensus, or to the demands of political and social power" (Ratzinger 2007).[177] Consequently, conscience furnishes proxies with reliable guidance for making decisions on behalf of the research subjects entrusted to their care.

In this context, it is astonishing that the human-rights documents scrutinized in the first part of this work never refer to conscience, with the exception of the Declaration of Helsinki, which mentions conscience only once and even then in a rather noncommittal manner, a shortcoming previously noted. No meaningful provision for counseling as an essential part of informed consent and no reference to its intimate link to formation of conscience are to be found in these documents. Nor can they be found in the approaches promoting self-determinative autonomy critically referenced throughout earlier chapters. Even the National Commission, whose stand on autonomy and the role it plays in informed consent seems more accurate, fails to recognize the significance of conscience in surrogate decision making.

The remarks offered by McInerny and Fisher clearly demonstrate that a well-formed conscience alone, despite its importance and indispensability, cannot ultimately protect the good of research subjects. Therefore, recourse to virtues, in particular the virtue of prudence, proves critical in achieving this protection. While conscience furnishes the proxy with knowledge of what good ought to be done, the virtue disposes the proxy actually to do this good. Thus, training in virtue proves no less important than formation of conscience. As is the case for all human beings in their daily lives, maturing in virtue conditions the moral quality of actions performed, and this is even more so for investigators, research subjects, and surrogates. I have earlier expressed my strong disagreement with the dominant logic undergirding various ethical codes and guidelines, which are based on an erroneous conviction that it is procedures and regulations implemented in research rather than the moral quality of those who implement them that play the most vital part in the otherwise commendable endeavor to render research with human subjects morally licit.

I have also pointed out the importance of virtue and moral character in both research in general and informed consent in particular. As Pellegrino and Thomasma observe, this importance is very characteristic of the field of biomedical ethics, "where rules, principles, and guidelines have dominated for the last quarter of a century." Consequently, "an ethic of virtue must complement the existing ethic of principles if we are to have a comprehensive perspective on the ethical behavior of the scientist" (Pellegrino and Thomasma 1993, 133).[178]

Needless to say, the same pertains to the behavior of the proxy. Yet, I would emphasize that implicit in this claim for compatibility between an ethics of virtue and an ethics of principle should be the conviction that these two types of ethics do not oppose one another. As a matter of fact, one acquires virtue by choosing to act in accord with the truths safeguarded by principles. The point is, however, that all too often in present-day biomedical ethics the latter occupies the distinguished position of the sole point of reference as if virtues were either nonexistent or not important at all. Furthermore, as is evident in light of the previously mentioned link between conscience and prudence, both the ethics of virtue and the ethics of principles must allow for a meaningful notion of conscience, including its formation. This in turn presupposes the existence of the objective and knowable good. Accordingly, I submit that the principles articulated in the human-rights documents need to be interpreted according to objective moral standards prior to their application. This task becomes even more urgent in face of the fact that the human-rights documents indicate no benchmark by which to evaluate their content.

Because the interpretation of moral principles and their application under concrete circumstances belong to the very nature of acting as proxy, the ethics of proxy decision making must account for both conscience and virtue as intimately linked to the objective good ingrained in human nature. It is noteworthy that an approach so construed fully satisfies the criterion of reasonableness, because, as already shown, the natural human inclination to act in conformity with reason means, among other things, acting according to virtue.[179] Thus, it refutes the content-thin, subjectivistic, and intuitive misunderstanding of reason permeating the standards of proxy decision making critically examined in this study. Finally, as previously pointed out, by saying that a proxy acts reasonably, in the objective sense of the word, one points to the fact that he or she, among other things, is habitually capable of avoiding emotional prejudice or bias. A failure to do so, in itself more likely to occur when the proxy belongs among the close friends or family members of the ward, undermines the validity of proxy consent, yet it can still be condoned in subjectivistic accounts, which are ill-prepared to cope with the problem of prejudice or bias.

11.2 Moral Absolutes and Human Dignity

All three levels of conscience referenced by the *Catechism of the Catholic Church* and noted in the previous section reflect the structure of natural law. Thus, the first level, synderesis, denotes "the habitual knowledge of the common precepts of natural law" (McInerny 1997, 103). St. Thomas recognizes these precepts as "imprinted on the natural reason to which they are self-evident" (ST, I-II, q. 100, a. 3), that is, "no man can have an erroneous judgment about them" (ST, I-II, q. 100, a. 11).

Two basic classes of self-evident precepts can be singled out. One class encompasses those precepts that "direct us to pursue the good and avoid its contrary (e.g., *good is to be done and pursued and its opposite, evil, is to be avoided*)" whereas the other regards "our way of pursuing such good (e.g., we are to do so by acting fairly [the Golden Rule], by refusing to do evil to anyone, and by loving God and neighbor)" (May 2003a, 77).

The second level of precepts builds upon synderesis and includes norms that are more detailed yet easily comprehensible even for the uneducated. Aquinas identifies these norms with those of the Decalogue.

Finally, the third level includes the precepts "the reason of which is not so evident to everyone, but only to the wise" (ST, I-II, q. 100, a. 11). And by "wise" Aquinas means the prudent or virtuous. Thus, the third-level precepts require "much consideration of the various circumstances" (ST, I-II, q. 100, a. 1) which in turn implies a well-formed conscience.

Now, I want to focus on precepts that specify certain kinds of acts, describable in non-morally evaluative language. They are commonly referred to as moral absolutes. These precepts correspond to the second and third level of natural law and hold universal and immutable standing as they proscribe intrinsically evil acts (John Paul II 1993, 115) to which they "admit of no dispensation whatsoever" (ST, I-II, q. 100, a. 8). In contradiction to this sound interpretation, certain revisionist theologians, such as Richard McCormick, deny the existence of moral absolutes and reduce their meaning to a merely formal aspect. According to McCormick, specific precepts of this kind express "useful generalizations but not universally true moral norms" (May 2003a, 170; see McCormick 1981a, 349–71, 529–44, 582–84; see also McCormick 1979, 1998, 2006, 58–59).[180] Hence, these precepts would allow for exceptions, from which it could be concluded that the actions they forbid are in fact not intrinsically evil. By saying this, i.e., by accepting as morally legitimate actions unconditionally enjoined by natural law, McCormick not only rejects the objective moral norms embedded in that law but likewise undermines the moral obligations that derive from them.

Ultimately, by dissolving the solid universal meaning of moral absolutes, McCormick attacks the goods they safeguard. Put simply, these goods are no longer inviolable because the norms protecting them now permit exceptions. Obviously, a non-objective norm cannot satisfactorily safeguard an objective good. This, I believe, is a momentous conclusion regarding the correct reading of McCormick's argument justifying proxy consent to research with children, which was discussed earlier in this study. By way of reminder, McCormick initially held that such consent is justifiable on the grounds of the presumption of the minor subjects' wishes, based on what they would do because they *ought to.* It is noteworthy that such reasoning makes sense only when predicated on an objective account of human obligation. Otherwise, one could never arrive at the knowledge of the child's duties because small children, and even more so infants, can neither grasp the meaning of any obligation nor voice any preferences of their own in this regard. Therefore, employment of the objective standard proves necessary to render McCormick's approach usable.

In fact, as noted above, McCormick himself explicitly referred to the objective standard, a fact, I believe, best seen in the account of children's sociality. There he argued that the child ought to want to pursue the good of others not because such pursuit relates somehow to his or her own good, but rather because it does not pose any real threat to it and yet promises a substantial benefit to others. One of course cannot question much in this thread of reasoning insofar as it presupposes the objective moral good of the child. It is far from evident, however, what kind of good McCormick had in mind in this instance. Given his rereading of the natural law, that is, given his rejection of the existence of moral absolutes articulated in morally neutral terms, McCormick implicitly denied the objectivity of the inviolable human goods, thus rendering them violable. This, however, lays bare a conflict inherent in his position. On the one hand, contrary to his own stated position, he deviates from the objective interpretation of the child's good and a fortiori the child's obligations, yet he offers no reliable yardstick by which to determine these obligations. Furthermore, since, for the reasons given above, neither the child's good nor his or her obligations can be satisfactorily ascertained on subjective grounds, McCormick's account ends up being self-defeating. Nevertheless, possibly aware of these difficulties, McCormick actually continues in the course of his argument to ascribe an objective quality to the good of the child and somehow manages to derive out of this the moral obligations which the child would choose because he ought to.

Clearly, McCormick's account of proxy decision making thus proves inconsistent and invites serious reservations. As such, it also undermines the role played by conscience not only in proxy decision making but indeed in the whole of human life. As John Paul II phrased it, conscience "formulates moral obligations in the light of the natural law" (John Paul II 1993, n. 59). The light of the natural law becomes dim in McCormick's account, which rules out moral absolutes. On the basis of which norms are the proxies now to make their decisions? Perhaps, conscious to some extent of this flaw in his argument, McCormick highlights the vital part to be played in research by the scientific community, especially as far as the determination of minimal risk is concerned. While the role of scientists in this determination should certainly be acknowledged, one could hardly see any reason to extend the competence of scientific expertise into the realm of ethical judgment. Put briefly, even with scientific determination of minimal risk at hand, how is one to ascertain that such risk in truth does not threaten the child's good without knowing in what this good actually consists? Presumably, one could simply assume that no threat exists, but even if plausible, this assumption nonetheless remains unsupported in McCormick's ethical framework, a substantial flaw. Furthermore, while operating on McCormick's premises, can one also assume that a slight increase over minimal risk does not threaten the child's good either, especially if this assumption rests on the suggestion of the scientific community that such an increase is negligible?

Of course, as asserted in the preceding section, genuinely scientific truth never opposes moral truth about human beings. McCormick, however, not only deprives moral truth of its objective meaning but also, and perhaps as a result of this, reduces it to scientific truth. Consequently, even if Ramsey's critique of McCormick's position charging him with imposing the adult world on the child and May's analysis showing that McCormick imputed moral agency to children were incorrect, which I do not think to be the case, one could still raise an objection, pointing to McCormick's misconception of natural law.

This objection also sheds light on one of the major lines of the National Commission's critique against McCormick, according to which his position must be rejected precisely because it is a natural law position. Besides its failure to distinguish between different approaches to natural law, the Commission also implied that because the natural law arguments have become subject to widespread criticism, the McCormick account, predicated on natural law, also cannot be accepted. Yet, since no logical connection exists between the amount of criticism and the validity of the position criticized, the Commission's inference turns out to be as baseless as it is erroneous. More significantly, as just explained, even as he strove to root his approach in objective universal norms, McCormick nonetheless denied the existence of such norms, thus placing his reasoning at variance with the account of natural law proposed by Thomas Aquinas and developed by the Magisterium. These accounts have consistently recognized the existence of moral absolutes and their role as safeguards of the objective human good.

By safeguarding the human good and by calling for its respect, moral absolutes protect "the inviolable personal dignity of every human being." In doing so, "they help to preserve the human social fabric and its proper and fruitful development" (John Paul II 1993, n. 97). In addition, "the dignity of the human person implies and requires *uprightness or moral conscience*" (*Catechism of the Catholic Church* 1997, n. 1780, emphasis in text). Put another way, obedience to the voice of conscience constitutes "the very dignity of man" (*Gaudium et Spes*, n. 16, after Flannery 2004).

As shown previously, informed consent, too, when properly understood, implies compliance with conscience. Although in itself not a moral absolute, informed consent intends to secure respect for persons and aims at the protection of their dignity. Early in this study, while addressing the principle of respect for persons articulated by the Belmont Report, I suggested that the interpretation and application of this principle should not be reduced to safeguarding the subject's autonomy. This demand became even more obvious when the profound inadequacy of the secular, content-thin notion of autonomy was demonstrated. In proxy decision making, where the autonomy of the subject is compromised or nonexistent, the need for an account predicated upon objective moral standards is most urgent.

Accordingly, I propose that instead of associating proxy consent with subjective and intuitive measures, the benchmark of inviolable human dignity guarded by moral absolutes must be employed. As a matter of fact, whether or not the research subject can exercise his or her autonomy does not enter into the equation at this level. Much more important is that as a human being the research subject is a person, radically capable of acting in a rational way, as a free, moral agent. In relation to this, Patrick Lee underscores that "the capacities to reason and make free choices are basic, natural capacities, possessed by every human being, even those who cannot immediately exercise these capacities." Moreover, "if dignity derives from the kind of thing one is, a human being—and surely this is dignity in the most important sense—then one cannot lose this dignity as long as one remains a human being" (Lee 2001, 341).

From the theological perspective, the universal and inalienable character of human dignity arises out of the truth of the creation of all men and women in God's image and likeness. Aquinas explains that man is *imago Dei*, the image of God, "by reason of his intellectual nature" and that this happens in three ways. First, at a most rudimentary level all human beings "possess a natural aptitude for understanding and loving God" which belongs to "the very nature of the mind." Second, humans become the image of God, insofar as they grow in actual and habitual knowledge and love of God, both of which, although imperfect, develop in conformity with grace. Third, "inasmuch as man knows and loves God perfectly" *imago Dei* "consists in the likeness of glory." Upon concluding these remarks, Aquinas notes that the first type of *imago Dei* is found in all persons, the second "only in the just" and the third "only in the blessed," that is, in the life to come (ST, I, q. 93, a.4).

Thus articulated, the theological perspective emphasizes that human dignity, considered in the second and third levels, is dynamic and allows for degrees, i.e., every human being can excel in it or diminish it even to the point of its loss. In other words, at the second level, we give dignity to ourselves with the help of God's infallible grace as we shape "our choices and actions in accord with the truth." The third level of dignity, in turn, is based upon God's gift of divine life, in which all are invited to participate (May 2000, 55). It is then up to the particular person whether or not he or she responds to this gift. Consequently, the realization of both second- and third- level dignity directly hinges upon the person.

In contrast, no human being, regardless of his or her spiritual, developmental, physical, or moral condition can ever lose dignity at level one,[181] a truth well articulated by Lee in the passage quoted above. Consequently, without excluding the other levels of dignity, this is the most fundamental level at which informed consent in the research setting must operate in order to attain the status of a reliable tool. By this I mean that informed consent must be a tool significantly more substantial than a merely intuitive device aimed at the protection of autonomy stripped of content. Such a limited device, rather than safeguarding the subjects as informed consent must do, often protects only investigators or proxies by, for example, shielding them from the legal consequences of their actions.

Human dignity, in particular its first level common to all humans, provides "the foundation and goal of the responsibility to protect," as Pope Benedict XVI emphasized recently during his meeting with the members of the General Assembly of the United Nations Organization (Benedict XVI 2008b). In his address, the Pontiff said, among other things, that the responsibility to protect must be seen "as an aspect of natural reason shared by all nations, and the result of an international order whose task it was to regulate relations between peoples. Now, as then, this principle has to invoke the idea of the person as image of the Creator, the desire for the absolute and the essence of freedom." Although enunciated in the context of the international order implying reciprocal cooperation among nations, the principle of the responsibility to protect, founded on human dignity and aiming at its protection, applies precisely to informed consent and comprehensively expresses its purpose. Put briefly, informed consent, including its proxy type, even though not a moral absolute, strives to protect the subject responsibly, i.e., in a way respectful of his or her humanity, naturally inclined as it is toward the objective good and truth.

Strikingly, human dignity goes unmentioned in the human-rights documents and in the accounts of proxy decision making addressed earlier. Nevertheless, all these documents and accounts, except for the Nuremberg Code, frequently resort to human rights. Correctly understood, however, human rights are embedded in natural law and thus maintain an inseparable bond with human dignity. To invoke Benedict XVI's wording, "the common good that human rights help to accomplish cannot... be attained merely by applying correct procedures, nor even less by achieving a balance between competing rights" (Benedict XVI 2008b). Yet a strong emphasis on procedures paired with an attempt to strike a balance between competing rights are the precise characteristics of both the human-rights documents, again with the exception of *Nuremberg*, and the accounts of proxy decision making examined previously.

Severed from the objective good to which they are supposed to give expression, human rights undergo a reduction to mere interests, minimal in content and indicative solely of subjective preferences. Consequently, human rights thus construed easily come into conflict with each other, and, indeed, without recourse to inviolable human good, these conflicts rarely appear resolvable. Here again, while aware of the conflicts in question, the human-rights documents and the accounts of proxy decision making fail to supply a meaningful remedy. The CIOMS Guidelines, for instance, discuss such conflicts only in the context of the duties proper to review committees. Thus, the commentary on Guideline 3 mandates that "*as far as possible*, there must be assurance that the review is independent and that there is no conflict of interest that might affect the judgment of members of the review committees in relation to any aspect of the research" (emphasis added). Yet without the objective underpinning, this regulation carries little, if any, meaning and hence is ill-equipped to protect the research subjects' dignity against the objectivization that arises when human rights are considered apart from human good. For example, what does the phrase "as far as possible" actually mean? What kind of possibility is it assumed here? Who and on what grounds is to define the boundaries of such possibility?

Furthermore, since it is incumbent upon the members of ethics committees, i.e., upon their own ethical convictions and insights, to determine the practical implications of human rights, the results are likely to depart from objective moral standards and simply reflect personal preferences. To be clear, personal in this context means private and hence subjective, if not subjectivistic, which in turn presupposes that different ethical boards will settle the same matters differently, a corollary of the absence of an objective yardstick.

Finally, while they put an exclusive emphasis on the ethics committees, the CIOMS Guidelines say no word on researchers and their duties, which, I would suggest, implies that the approval of a research project by the ethics committee can absolve the researchers' moral responsibility. Thus, not only is the role of informed or proxy consent diminished but the principle of responsibility to protect, rooted in human nature and aiming at the protection of human dignity, in itself indispensable to both investigator and proxy, is dismissed.

The perspective of "responsibility to protect" can be said to introduce an objective moral quality into the informed-consent process. Yet, while defending against downgrading consent to a merely civic or punitive device by stressing its ethical character, this responsibility also points to a holistic anthropology that "respects the whole person" (John Paul II 1999). In other words, to be responsible as a proxy for the research subject means to protect this subject's integrity, not just some of the rights or selected aspects of the subject's well-being. It means to protect the whole good of that subject in its integrity. Respect for the whole person demands that no person may ever be reduced only to one part or only to a sum of his or her parts. In addition, in no aspect, be it physical, psychological, or spiritual, can any person be objectified, i.e., treated as a means only, in subordination to the interests of others. Therefore the Principles and Guidelines correctly pronounce that the subject is "the most important aspect of every clinical trial," as opposed to any "procedure, process, or product being studied" (Guideline 3c). Obviously, the subject's significance reaches far beyond any interest or profit others may gain from the subject's participation in research.

11.3 Proxy Consent and the Subject's Good

The analyses contained in the preceding two sections prove indispensable for a comprehensive understanding of informed consent, including its proxy type. Rescued from the trap of subjectivism, informed consent retains its meaningful role in both treatment and research as rooted in and expressive of objective morality, and hence capable of safeguarding human subjects or patients. As such it also provides reliable guidance for proxies who, in case of doubts of conscience or conflicts of interest, find in objective morality concrete indications of what is morally good and what is not, that is, what needs to be done in a given set of circumstances and what must be avoided.

With the objective yardstick at hand the very definition of informed consent also undergoes transformation as it now respects both the human good and the truth. In present-day bioethics literature the requirement of *free* and *informed* consent is typically understood as stipulating that *reasonably free* and *adequately informed* consent be procured. Certainly, as discussed in the second chapter of this study, securing fully free and informed consent is rarely if ever feasible in real-life research settings. Therefore, for example, the deluge of information provided to research subjects, although having the noble aim of fully informing the subjects, thwarts rather than augments the consent process, which inevitably diminishes its quality.

Regarding the "reasonable freedom" criterion, it unfortunately is reduced all too often to pure self-determinative autonomy indifferent to the natural inclinations towards the good and truth, the problem discussed throughout Chapters 3 and 4. Within this flawed perspective the faculty of reason and its role in the moral fabric of human life is radically and arbitrarily diminished, a process fraught with consequences. As already demonstrated, in modern accounts of surrogate decision making reason becomes an enemy of the indifferent will and its self-determination, which is autonomy. In effect, with all the power of choice reserved for the will, reason not only ceases to present the good to the will but simply is removed from the realm of moral choice. Despite that, however, it is still invoked in these accounts, most often under the guise of the so-called reasonable-person standard. The problem is that, as shown in Chapter 3, this standard, devoid of objective content, can do no better than to equate reasonableness with social consensus, by virtue of which the determination of what counts as reasonable is left to the autonomous decisions of members of society. Yet, divorced from the objective good that the faculty of reason presents to the will, these decisions are essentially arbitrary, shaped by the predominant existing culture. Therefore, insisting in this context that informed consent is to be *reasonably free* means no more than treating it as a subjective device, which, as it is alienated from human good, cannot protect this good either in treatment or in the research milieu, and thus does not bear scrutiny. To become workable, reasonably free consent must presuppose freedom of excellence along with the objective account of reason as unified with the will. Only in such a way are conscience and virtue meaningfully preserved and human dignity satisfactorily protected.

In the same vein, the second characteristic of informed consent, that is, "adequate information," calls for re-description. In modern accounts of surrogate decision making examined in Chapter 3, the phrase *adequately informed* points to an informational balance tailored to the subjective needs of the research subject. Put otherwise, from the set of relevant information a subset of material information is to be singled out so as to meet the subjective needs of a given subject. No other, more reliable, criterion for the selection of material information is offered. Moreover, in the communication process thus perceived, only the scientific truth is relayed, without any reference to the moral truth that demands protection of the subject's human dignity. As is evident at this point, adequate information calls for employing both scientific and moral truth. This in turn requires a genuine dialogue between the investigator and the subject, not one reduced to a mere exchange of technical data finalized by securing the latter's signature on the consent form. The dialogue in question implies sincerely facing the reality of the research project's implications (both technical and moral) as well as the subjects' condition (again, both medical and moral). As explained in Chapter 2, such dialogue allows for counseling, which presupposes and not infrequently also aims at the formation of conscience.

Undoubtedly, a dialogue meeting the objective moral standards also needs to take place between the subject and the proxy before the subject slips into incompetence. In case of the never-competent subjects, such as infants, the dialogue is still sustained by the investigator and the proxy with its purpose being to act in conformity with the ward's objective good. Obviously, conscience and prudence have a critical part to play in such a dialogue. Even though in therapeutic research the practical determination of what is adequate and reasonable comes about by employing the Golden Rule, as discussed at length in Chapter 3, resorting to the Golden Rule may turn out to be insufficient in nontherapeutic settings. In other words, although meaningful exercise of the Golden Rule always presupposes the objective good, insofar as proxy decision making regarding nontherapeutic research is concerned, a more precise determination about the amount of risk that violates the subject's good must be made. Such determination is of utmost relevance because a higher degree of risk in nontherapeutic settings cannot be justified by the favorable ratio of beneficial outcomes to the subject, as often happens in therapeutic situations. Chapter 4 demonstrated how complex, and even convoluted, the lines of reasoning set forth to deal with this dilemma can be. Certain authors, such as Ramsey, see any amount of risk as irreconcilable with the subject's good and accordingly pronounce a total ban on nontherapeutic research with small children and by extension also with other incompetent subjects, regardless of the risk involved. Other authorities, however, including the Magisterium and such thinkers as Germain Grisez, the later William May, Benedict Ashley, and Kevin O'Rourke find this interpretation too stringent because some nontherapeutic procedures, for example those merely observational in nature, clearly cause no harm to the subject, which is to say that they do not put his or her inviolable good in jeopardy. In my view, under the additional condition that such procedures hold a promise of considerable benefit to others, it is indeed impossible to see any reason to judge them morally illicit.

It is also worth noting that while virtually all current authors who support this conclusion consider minimal risk as its prerequisite, the Magisterium exerts more caution in this respect. Rather than employing the standard of minimal risk, it speaks instead of the absence of significant risk. This approach, as demonstrated in the last section of Chapter 4, essentially rules out any research involving more than minimal risk as well. Yet the Magisterium's wording is superior in that it does not engage in the controversy surrounding the minimal-risk criterion and its definition. Putting the matter in terms of the lack of significant risk proves relevant also because even if deemed minimal medically (i.e., at the physical level), the risk involved in a given research endeavor can still hold harmful potential to the subject at the emotional or psychological level, a fact not necessarily always known to investigators. Finally, by not conditioning the moral justifiability on the minimal-risk standard predominant today in professional literature, or on any other external benchmark, the Magisterium calls attention to the fact that ultimately it is a matter of both a well-formed conscience and of prudence on the part of investigators, proxies, and other individuals participating in the decision making process to arrive at the certainty that the nontherapeutic research in question does not in fact threaten the subject's good. For this reason, the National Commission's endorsement of justifiability of the so-called "slight increase over minimal risk" referenced in Chapter 4 cannot be taken for granted. Put briefly, it may indeed occur under some circumstances that such an increase will not jeopardize the subject's good, but in order to determine whether or not this in truth is the case one needs to resort to a well-formed conscience and must use prudence. Regrettably, the Commission includes neither conscience nor prudence in its considerations. Nor does it indicate the way in which the slight increase over minimal risk should be ascertained so as to respect the good of the incompetent subjects.

From what has been said thus far it follows that it is the subject's inviolable good that constitutes the basis for the justifiability of nontherapeutic research with incompetent subjects, which means that proxies can licitly permit such research only when the objective good of their wards is safeguarded. A practical way of assuring that the latter indeed is the case consists in verifying that no significant risks are at stake. Building upon these fundamental premises, I wish to propose another rationale underpinning the moral validity of the subjection of an incompetent person to nontherapeutic research, when such research does not violate that person's good. The rationale draws from the sociality natural to all human beings and has to do with the close relationship between the individual good and the common good, in this case represented by generalizable medical knowledge to be gained from nontherapeutic research.

The Magisterium views the common good as "the sum total of social conditions which allow people, either as groups or as individuals, to reach their fulfillment more fully and more easily" (*Gaudium et Spes*, n. 26, after Flannery 2004).[182] Implicit in this definition is the truth that human beings, by virtue of their very nature, are social, i.e., they need others to achieve fulfillment. Put another way, no human person can "find fulfillment in himself, that is, apart from the fact that he exists 'with' others and 'for' others" (Pontifical Council for Justice and Peace 2004, n. 165). Accordingly, the *Catechism* asserts that "the common good concerns the life of *all*" (n. 1906, emphasis added), which is to say that it concerns both children and adults, both moral agents and those humans with diminished or nonexistent moral agency. From the fact of their sociality, it follows that *all* human beings have a share in the common good. Indeed, properly understood, the individual good and the common good, although distinct, are mutually exclusive only to a degree (Smith 1995, 164). To help clarify this matter, certain authors distinguish, after Aquinas, two basic kinds of *bonum suum* (one's own good). First is the good that is both our own and others' which is to say that while being properly ours this good is also shared with others. Such is precisely the nature of the common good which is shared by all.[183] The second type of *bonum suum* denotes the private good, i.e., one that belongs only to us and that cannot be shared with others (Smith 1995, 87).

Two important points need to be made in the light of the description of the common good just offered. First, the common good refers not merely to an aggregate of the private goods or the sum of the goods shared only within certain social groups. Rather, "belonging to everyone and to each person," this good, "is and remains 'common,' because it is indivisible and because only together is it possible to attain it, increase it and safeguard its effectiveness" (Pontifical Council for Justice and Peace 2004, n. 164). Needless to say, the common good, thus construed, precludes any totalitarian definition because the government of the state, as of any other community, has the same duty of serving the common good as do the citizens. In the totalitarian setting, in contrast, the common good is perverted and is reduced to the pursuit of the ideological interests of those in power at the cost of the violation of the good and the dignity of those ruled.

The second major point is that since the common good, correctly understood, signifies the good shared, and since there is a fundamental natural inclination to live in community, it follows that the goods that we share are more important to us than those that we cannot share. Put otherwise, given that each true good is "perfective and fulfilling of the agent" (McInerny 1997, 2), the goods we share are more perfective of us than are private, non-shareable goods.[184] In this perspective and in terms of one's goals and values, it is clear that the pursuit of the common good ought to take precedence over the pursuit of the private goods. Still, sharing the common good by virtue of being human does not amount to pursuing that good. Certainly, while all human beings share in the common good, only moral agents are capable of pursuing it. In consequence, even though infants and other incompetent research subjects cannot pursue the common good, they still have their share in it and indeed contribute to it, most fundamentally by their very existence.

With these distinctions in mind, let us now raise the question: can parents in the name of pursuing the common good subject their infant to nontherapeutic research? The answer first depends on whether or not the research in question *truly* serves the common good. Clearly it does when two conditions are simultaneously satisfied: (1) its design is scientifically sound, i.e., it enhances generalizable knowledge; (2) it does not threaten the child's inviolable good. If the research did pose a threat to the child's good, it would thereby menace the common good as well, because, as the *Catechism* teaches, "the good of each individual is necessarily related to the common good, which in turn can be defined only in reference to the human person" (n. 1905). For this reason, the common good always "presupposes *respect for the person* as such" (n. 1907). Such respect demands that no significant risk be present, which is to say that the risk involved does not harm the subject. Yet this provision, as already discussed in the fourth chapter of this study, does not apply to the unborn human subjects whose good may indeed be endangered by *any* active intervention including that featuring a degree of risk not exceeding minimal. Moreover, in line with May's explanation, also referenced in Chapter 4, respect for persons requires that on account of their great vulnerability, infants and other incompetent individuals never be enrolled in nontherapeutic research unless (1) given the research objectives, it is not possible for less vulnerable subjects to participate instead; (2) the research features only slight inconvenience and (3) the expected benefits are significant.

When all the requirements enumerated above are met, I believe that the answer to the question opening the previous paragraph is positive. The proxy, be it the parent or legal guardian, can submit an infant to nontherapeutic research satisfying the above-listed conditions because such research contributes to the common good in which the infant shares by virtue of his or her humanity. Correctly understood, the common good not only fully respects the good of the infant but also, since it is shared, is in fact part of the infant's own good. This conclusion brings us back to Ashley's position quoted in the previous chapter, in line with which every child, and indeed every human being, constantly benefits from sharing with others. It is worth noting that this observation holds true also in research settings. In real-life situations, when the infant is subjected to nontherapeutic research, with the proviso that given the research's objectives it is not possible to recruit less vulnerable subjects instead, such research virtually always aims at the generalizable knowledge which is relevant to the class of infants. For this reason, under some circumstances it may turn out that the infant subject will actually benefit from such research in the future. Even if he or she does not, the participation in this research will surely contribute to the common good of the class of infants. In this context, it needs to be emphasized that every infant, from birth or even prior to birth, profits from medical care which in turn presupposes knowledge gained by way of research. But even at a more rudimentary level, as Ashley suggested, in the wake of his participation in scientifically sound and morally legitimate research every infant will benefit from belonging to a society rendered healthier by means of such research.

Before completing this section, I must note contra Ramsey that parents subjecting their infant children to nontherapeutic research do not treat them as a means only, as long as they ensure that the child's good is protected. Therefore, even if not directly enhancing the child's individual good, nontherapeutic research can still contribute to this good by virtue of its being at the service of common good, which, as just explained, translates into the child's own good as well.

Finally, it is in order to emphasize that allowing for participation in such research meets the criteria of a morally good act. Provided that the qualifications listed above are met, the object of the act, i.e. of enrolling the infant in nontherapeutic research involving no significant risk is morally good because it respects the good and the dignity of the research subject. The intention, of the proxies is good, too, as long as they act for the sake of the common good. If the intention were wrong, i.e., severed from the common good and thus also from the good of the infant, one could, I believe, speak of the "offensive touching," in the sense that Ramsey attached to this term. Last but not least, the circumstances of the act in question themselves are not capable of radically altering its moral quality so long as the requirement of no significant risk is met. Still, as already noted, additional requirements proper to the level of circumstances, such as the absence of major inconvenience for the research subject or the impossibility, caused by the research objectives, for less vulnerable subjects to be recruited instead, although not constitutive of the morality of the act under review, nonetheless play a significant part in a broader justificatory framework by aiming at an even greater protection of the incompetent subject in nontherapeutic research.

Conclusion

What is the nature of proxy consent to experimentation on human subjects and what are its limits? Having answered these questions over the course of the preceding chapters, it is now appropriate to summarize and recapitulate the major points of those findings here.

Regarding the nature of proxy consent, it is important to begin by noting that, correctly understood, proxy consent designates neither the extended nor the presumed consent of the subject. Rather, it is the personal consent of a proxy, expressive of his/her responsibility for an incompetent ward. In contrast to the accounts of surrogate decision making predominant in present-day bioethics literature, the responsibility in question does not amount merely to an autonomous exercise of rights on the part of the proxy. Likewise, it cannot be reduced to autonomy perceived as pure self-determination.

Even though commonly endorsed by mainstream bioethicists as a primary rationale for proxy consent, the principle of (respect for) autonomy is fatally flawed. Devoid of content, it presupposes an erroneous anthropology that denigrates reason to the point of its radical renunciation. In brief, the faculty of reason is not recognized as presenting the good to the will and the will itself is construed as if it could not will the good. Thus, reason and will become enemies. Divorced from the objective good, the will chooses arbitrarily among available options. The will, inimical to reason, effectively ends up taking over the role of reason. In this system, a voluntaristic account of human freedom obfuscates rather than explicates, and ultimately undermines, proxy consent. Indeed, it neither safeguards the good of the research subject nor provides reliable guidance for proxies. Nonetheless, the split between will and reason plays a vital role in the mainstream accounts of surrogate decision making. Alas, the human-rights documents discussed in this study, with the exception of the Nuremberg Code and (to a lesser degree) the Belmont Report, fail to avoid the trap of arbitrariness and focus—out of concern for securing autonomy—on the self-expression of the research subject (even if merely caprice or whim) rather than on his or her real good. In contrast, proxy consent, accurately understood, implies a responsibility aimed at the objective good of the ward. As such, it finds its rationale in the Golden Rule, which is rooted in a meaningful account of the good and thus recognizes the basic inclinations natural to all human beings, including the need for friendship and love.

G. Mazur, O.P., *Informed Consent, Proxy Consent, and Catholic Bioethics*, Philosophy and Medicine 112, DOI 10.1007/978-94-007-2196-8,

Because knowledge of the good conditions the very possibility of freedom, any attempt to predicate proxy consent on purely self-determinative autonomy that dismisses the good necessarily undermines freedom. Therefore I regard it as critical to emphasize that it is freedom rather than autonomy that underlies proxy consent. Phrased another way, proxy consent presupposes and is expressive of free choice rather than autonomous choice. Constitutive of free choice is an unqualified resistance to any kind of indifference toward goodness and truth. Such, indeed, is the freedom that belongs to proxy consent's very nature.

Contingent upon the nature of proxy consent are its limits. These limits are fundamentally conditioned by the good of the incompetent subject. Consequently, proxies are entitled to consent only to what respects the good of their wards. Although this is not always immediately evident, the good in question is discovered or rediscovered by means of the trustful dialogue of the proxy with the researcher and the subject's family members or friends. Needless to say, such dialogue becomes even more important when the proxy does not know the ward well. Apart from this, consent always requires a well-formed conscience and virtue, both on the part of the proxy and those otherwise involved in the process of decision making.

Safeguarding the objective good of the subject does not preclude taking into consideration the subject's preferences expressed before slipping into incompetence. By the same token, whenever feasible, the assent of the subject should always be sought as well. This includes child subjects who, although not sufficiently mature to offer full-blown informed consent, frequently prove capable of cooperating meaningfully in the process of decision making with respect to their participation in research.

It is worth noting that a sound concept of objective good rooted in human nature protects more than just those individuals who serve as research subjects. It also safeguards the proper role of researchers themselves. Indeed, without such a concept, researchers can easily be reduced to mere technicians with their duties restricted to supplying medical services at the patient's or subject's wishes, without even knowing whether this will have a beneficial outcome apart from the purely medical one. Obviously, a good medical outcome is neither the only dimension of the good of the human subject nor a necessarily overriding one. Therefore, the medical good cannot by itself determine the morality of proxy consent. Rather, the moral good of the subject has to be considered in its totality. This is to say that scientific truth must neither be separated from the moral truth about human beings nor be treated as superior to it. The importance of moral truth pertains primarily to human dignity which in turn requires that no human being be treated solely as a means but also as an end.

It is at the juncture between human dignity and the moral truth about the subject (i.e., the truth about the subject's objective good) that a potential quandary regarding the morality of nontherapeutic research and its concomitant risk emerges. In short, is it licit to subject the incompetent person, such as infant, to nontherapeutic experimentation? The answer to this question, as proposed in this study in keeping with the Magisterial teaching, is positive, provided the risk involved is not classified as significant.

Insignificant risk, in turn, is to be understood as a degree of risk which does not jeopardize the good of the subject. Furthermore, two additional points need to be made. First, with the subject's good being protected, nontherapeutic research must contribute to the common good, which is to say, it must be soundly designed scientifically. Second, all humans by reason of their sociality share in the common good; in other words, the common good, properly understood, becomes their own good as well. Therefore I conclude that it is morally licit to subject an incompetent person to nontherapeutic research provided that no significant risk is involved.

In order to respect the subject's objective good, a practical evaluation of whether the degree of risk is insignificant requires, besides taking into account scientific data, the knowledge of the research subject as well as the use of conscience and prudence in acting. This perspective stands in stark contrast to the minimal-risk standard, endorsed in today's mainstream bioethics literature, while also avoiding the controversy surrounding the minimal-risk criterion's actual meaning and scope.

Notes

1. As stated on the title page, this book has been written with the assistance of Alexander Morgan Capron and Eleanor Swift Glass.
2. As explained in "Acknowledgments" Nancy King played the role of a collaborator and legal consultant to Faden and Beauchamp in the course of production of this book. She also authored chapter four entitled: "Consent and the Courts: The Emergence of the Legal Doctrine." Because I make no use of this chapter in the analyses contained in this study, in the body of the text I refer only to the primary authors, i.e., Faden and Beauchamp.
3. Pernick's personal communication with Faden and Beauchamp (quoted in Faden, Beauchamp, and King 1986, 58).
4. On both Percival's education as well as on the substantial influence some clergymen and theologians had on his personality and intellectual legacy see Haakonssen (1997, 94–110).
5. For more on this see Pellegrino and Thomasma (1993, 33–34).
6. It was not only the Nazis, however, who during World War II proved that what philosopher Hanna Arendt calls "the banality of evil" can under certain circumstances come about so easily. The engagement by Japanese physicians in horrendous human experimentation has been recalled in Kimura (1998).
7. For a description of each kind see Taylor (1992); Spitz (2005, 65–233). Shocking coverage of some of the crimes performed by Josef Mengele in the Auschwitz and Birkenau concentration camps can be found in Pence (2004, 270–74). See also Lifton (2000, 337–83); Baumslag (2005).
8. "The lack of a defined ethical or regulatory code that governed the research practices of U.S. scientists caused some anxiety on the part of the American prosecutors. To help them in their efforts to understand the technical aspects of the case and the norms that governed international scientific practice, the prosecutors, at the direction of the American Medical Association (AMA), turned to Dr. Andrew Ivy. Dr. Ivy was a well known physician and researcher who had been involved in experiments similar to the desalinization and high altitude experiments carried out by the Nazi defendants" (Winslade and Krause 1998, 143).
9. Interestingly, on the same page, Grodin invokes two cases which call into question Bernard's faithfulness to his own conclusions banning any nontherapeutic research.
10. Hereafter I quote the text of the Nuremberg Code published in Jonsen, Veatch, and Walters (1998, 11–12).
11. I will return to this in a more detailed fashion in later sections of this study, particularly on the occasion of examining the Belmont Report and its principle of respect for persons.
12. "The tribunal judgment will be of profound and enduring value in the field of medical jurisprudence; and the trial as a whole is an epochal step in the evolution of forensic medicine. The trial illustrates, furthermore, how rapidly the focus of activity in international law has moved from the academic lecture hall and toward the courtroom. The Nuremberg proceedings are among the outstanding examples of modern international law in action." These words of the post-World War II chief U.S. prosecutor Telford Taylor are quoted in

Mitscherlich and Mielke (1949, xxvi). Also Grodin quotes these words in Grodin (1992, 140).

13. Some commentators, like Herranz, claim that "the Nuremberg Code's influence on ethical codes in medicine was late in coming and only partial." Moreover, "the Code was practically ignored by both national and international medical organizations from 1947 to 1975" (Herranz 1998, 127). Despite his criticism, Herranz does not deny that the Code became influential via Declaration of Helsinki that included almost all its provisions and "was the impetus for the Nuremberg Code's entrance into codes of various national medical associations" (Herranz 1998, 134).
14. This text has also been published as Appendix 5 in Annas and Grodin (1992, 346–52).
15. Hereafter I quote the texts of the Declaration from its first version (Helsinki I) through the fourth (Helsinki IV) published in Annas and Grodin (1992, 331–42).
16. Perley in fact adds one more alleged similarity by declaring that both documents "mandate that the foreseeable benefits of the research must be balanced against the inherent risks to the research subject." I chose not to include this statement because of its too generous approach to the Nuremberg Code. In fact, along the lines of the quote from Glantz invoked in the previous section, Nuremberg does not explicitly refer to the risk-benefits balancing.
17. A succinct and competent summary of the points in which the Declaration of Helsinki departs from the Nuremberg Code can be found in Annas (1992, 204–206).
18. The 2000 Declaration of Helsinki (Helsinki VI) supplements this with the following: "if the consent cannot be obtained in writing, the non-written consent must be formally documented and witnessed" (principle 22). Hereafter I quote the text of Helsinki VI published on the official website of the World Medical Association (see World Medical Association 1964).
19. The Declaration was first amended by the 29th WMA General Assembly, Tokyo, Japan, October 1975; then by the 35th WMA General Assembly, Venice, Italy, October 1983; for the third time, by the 41st WMA General Assembly, Hong Kong, September 1989; again by the 48th WMA General Assembly, Somerset West, Republic of South Africa, October 1996; and finally for a fifth time by the 52nd WMA General Assembly, Edinburgh, Scotland, October 2000. In addition, two Notes of Clarification were added by the WMA General Assembly, the first on Paragraph 29 in Washington in 2002, and the second on Paragraph 30 in Tokyo in 2004.
20. Some of the most outrageous examples from the recent U.S. history include the Tuskegee Syphilis Study (1932–1972), the Jewish Hospital Cancer Study (1963), the Willowbrook Hepatitis Study (1965), and the Krieger Lead Paint Study (mid-1990s). See Pence (2004, 276–300); Faden, Beauchamp, and King (1986, 161–67); Ramsey (2002, 47–58); Emanuel et al. (2003, 11–23).
21. For instance, 1989 Declaration of Helsinki (Helsinki IV) states that "the design and performance of each experimental procedure involving human subjects should be clearly formulated in an experimental protocol which should be transmitted for consideration, comment and guidance to a specially appointed committee independent of the investigator and the sponsor provided that this independent committee is in conformity with the laws and regulation of the country in which the research experiment is performed" (principle I.2). Helsinki VI adds that "the committee has the right to monitor ongoing trials. The researcher has the obligation to provide monitoring information to the committee, especially any serious adverse events. The researcher should also submit to the committee, for review, information regarding funding, sponsors, institutional affiliations, other potential conflicts of interest and incentives for subjects" (principle 13).
22. It is only the 2000 Declaration of Helsinki (Helsinki VI) that presents us with an attempt to clarify and specify a consent-waiver provision. Its principle 26 declares that "research on individuals from whom it is not possible to obtain consent, including proxy or advance consent, should be done only if the physical/mental condition that prevents obtaining informed consent is a necessary characteristic of the research population. The specific reasons for involving research subjects with a condition that renders them unable to give informed

consent should be stated in the experimental protocol for consideration and approval of the review committee. The protocol should state that consent to remain in the research should be obtained as soon as possible from the individual or a legally authorized surrogate."

23. It is not an easy task to define the term "placebo." According to Webster's Third International Dictionary, placebo is "an inert medicament or preparation given for its psychological effect especially to satisfy the patient or to act as a control in an experimental series" ("Placebo" 2002). Webster's Medical Dictionary offers a similar, although more detailed, twofold definition of placebo: (1) "a usually pharmacologically inert preparation prescribed more for the mental relief of the patient than for its actual effect on a disorder;" (2) "an inert or innocuous substance used especially in controlled experiments testing the efficacy of another substance (as a drug)." [*Webster's Medical Dictionary, Unabridged*, s.v. "Placebo," http://unabridged.merriam-webster.com (accessed August 8, 2008)]. The above definitions, although straightforward and commonsensical, contain, according to some thinkers such as Brody, "serious inconsistencies." The example introduced to support this claim refers to the term "inert remedy." "If the placebo was totally inert—Brody argues—there would be no point in giving it." (Brody 2004, 2030). I think this remark indicates some confusion that must be sorted out. First, none of the definitions explicitly speaks of a totally inert material but only of an inert one. Second, a placebo itself, as an inert substance, does not impact the subject's condition, in accordance with the very meaning of the word "inert." What affects the subject's condition is rather his or her positive attitude and faith in the effectiveness of the supposed medicine received. This remains true even though—as Brody points out—totally pure placebos hardly exist, which is to say most placebos currently in use have "pharmacological potency," at least under certain conditions. Still, it is not a placebo but rather the use of it that can never be judged entirely inert. A well-known phenomenon, directly caused by the use of placebo and labeled the "placebo effect," can be depicted as "an apparent improvement in health due not to any treatment but only to the patient's belief that he or she will improve (as by taking a dummy pill that is thought to be a cure). An opposite, or 'negative placebo effect,' has been observed when patients believe their health will get worse" ("Placebo" 2000) Along the lines of Brody's explanation, "for purposes of ethical analysis, placebo effect may be defined generally as the change in a patient's condition that results from the symbolic aspects of the encounter with a healer or with a healing setting, and not from the pharmacological or physiological properties of any remedy used." (Brody 2004, 2030.) In light of a good deal of bewilderment about the term "placebo effect," Hoffman et al. propose a distinction between the term "placebo effect" and another term "placebo response," with the purpose of rendering the terminology more clear and precise. In this approach, the placebo effect refers to "any average improvement in the condition of a group of subjects that has received a placebo treatment. It is an inference based on the assumption that, had the placebo not been given, no such improvement would have been observed." On the other hand, the placebo response relates to "the change in an individual caused by a placebo manipulation. Clearly, it is the placebo response—its psychology and neurobiology—that is most interesting to researchers today" (Hoffman, Harrington, and Fields 2005). For a critical review presenting various definitions of placebo effect and their inadequacies see Gotzsche (1994). To learn more on psychological dimensions of placebo effect and its connection with faith, see Miller (2005). A valuable article on placebo effect, offering a review of its history and the current state of research as well as a useful summary of current literature on the matter is Barrett et al. (2006). For a concise summary of some of the ethical issues, mostly those which are deception-related, regarding placebo and the placebo effect, see Brody (2004, 2031).

24. Some issues may be raised in regard to the phrase "inert placebo". Taking into account common definitions, this expression seems redundant because placebo per se constitutes an inert substance. Accordingly, the 2000 version of the Declaration discontinues using the term "inert placebo" in favor of the simple "placebo" (see Helsinki VI, paragraph 29). On other revisions adopted by Helsinki V, see Tollman (2001, 1418). Hereafter I quote the text of the fifth version of *Helsinki* (World Medical Association 1996).

25. For more information on this and some other revisions approved by Helsinki V, see Tollman (2001).
26. An informative description of these as well as other trials can be found in Kent et al. (2004).
27. An example is the antiretroviral drug Zidovudine that reduces the peril of maternal-infant HIV transmission by approximately two-thirds. See Connor et al. (1994); Sperling et al. (1996); Sia et al. (2004); Turner et al. (1999); Stringer and Rouse (1999).
28. For instance, the placebo-controlled trials performed in the 1990s in sub-Saharan Africa were sponsored by the National Institutes of Health (NIH) and the Centers for Disease Control and Prevention (CDC). For more details see Angell (1997).
29. On the gaps and shortcomings with respect to the informed-consent process and protection of confidentiality of subjects in developing countries see e.g. Hyder et al. (2004).
30. For a detailed presentation of the positions involved in the debate over this issue see Macklin (2004).
31. "We are anxious to ensure that no good ethical research is restricted, while at the same time we are adamant not to compromise the ethical principles the medical profession stands for" (Ferriman 2001, 1142).
32. The quotations of Note of Clarification on Paragraph 29 come from the full text of Helsinki VI available on official WMA website (see World Medical Association 1964).
33. Note of Clarification on Paragraph 30 as available on official WMA website.
34. For a detailed list of examples see Human and Fluss (2001).
35. Hereafter I quote the text of the Guidelines as published in Council for International Organizations of Medical Sciences (2002).
36. Guideline 8 deals with the benefits and risks issues in a detailed fashion. Right at the outset, the assertion is made that "for all the biomedical research involving human subjects, the investigator must ensure that potential benefits and risks are reasonably balanced and risks are minimized." Some procedures utilized in the course of research project benefit subjects directly, while others aim purely at producing new knowledge for science and society. In line with Guideline 8, "beneficial interventions are justified as they are in medical practice by the expectation that they will be at least as advantageous to the individuals concerned, in the light of both risks and benefits, as any available alternative. Non-beneficial interventions are assessed differently; they may be justified only by appeal to the knowledge to be gained. In assessing the risks and benefits that a protocol presents to a population it is appropriate to consider the harm that could result from forgoing the research."
37. The confidentiality-related issues have also been addressed by the Guideline 18 confirming that "the investigator must establish secure safeguards of the confidentiality of subjects' research data." Guideline 18 commentary has been divided into three sections. The first section refers to confidentiality between investigator and subject with a special emphasis on HIV/AIDS drug and vaccine trials that "may impose upon the research subjects significant associated risks of social discrimination and harm." The second section deals with confidentiality between physician and patient. By introducing a distinction between physicians and investigators, the commentary affirms that "a treating physician should not disclose any identifying information about patients to an investigator unless each patient has given consent to such disclosure and unless an ethical review committee has approved such disclosure." The third section of the commentary relates to the issues of confidentiality in genetics research. It suggests that "investigators must see that biological samples are fully anonymized and unlinked; this ensures that no information about specific individuals can be derived from such research or passed back to them."
38. In general, the Guidelines approach the issue in question in three major steps. First, they grant that securing informed consent is a process rather than a single event. In other words, signing the consent form by a prospective subject does not suffice to demonstrate that a genuine consent has thus been gained. Putting the name on the consent form should be an element of the entire process and yet neither the final nor an irreversible one. As stated above, the subject can withdraw consent at any time. Moreover, in long-term studies the

consent should be renewed "at pre-determined intervals, even if there are no changes in the design or objectives of the research" (Guideline 6). Renewal helps to reconfirm that subject understands the research procedures he is exposed to. Second, the attention must be paid to the language used in the course of conveying the information to the subjects. This "must not be simply a ritual recitation of the contents of a written document." Not infrequently, a substantial amount of patience and sensitivity need to be employed on the part of the researcher in order to suit the information to the subjects' comprehensive capacities. On occasion, it is advisable for a researcher to "administer an oral or a written test or otherwise determine whether the information has been adequately understood" (commentary on Guideline 4). Third, out of concern for guaranteeing full comprehension on the part of potential subjects, the Guidelines focus attention on cultural idiosyncrasies. As is well known, "in some populations the use of a number of local languages may complicate the communication of information to potential subjects and the ability of an investigator to ensure that they truly understand it." Moreover, in many indigenous settings the locals are not always familiar with medical concepts such as randomized clinical trial or placebo. This poses further obligations on investigators to overcome linguistic and cultural difficulties, which may result in more time and money spent on education of and informative dialogue with prospective subjects (commentary on Guideline 4).

39. Nevertheless, some of them have been also criticized for being too detailed and overloaded. For more on this, see e.g. Bhutta (2004, 774).
40. It must be admitted, however, that it is not true that the research was totally unregulated before Beecher's expose. In fact, the federal regulations can be traced back to 1962. In addition, some "regulatory exceptions" to largely unregulated research practices occurred as early as the 1950s; one example is NIH-sponsored research at the Clinical Center opened in 1953. For more on this, see Faden, Beauchamp, and King (1986, 200–14).
41. For more on this, see Faden, Beauchamp, and King (1986, 161–64). See also Sherwin (2005, 151).
42. As a matter of fact, there are many more issues with Tuskegee Studies. One of them pertains to the highly questionable scientific value of research done in unethical way on non-consenting subjects. See Faden, Beauchamp, and King (1986, 165–67); Pence (2004, 280–94); Vanderpool (1996, 9–10).
43. Hereafter I quote the text of the Report published on the official website of the National Institute of Health (see National Commission for the Protection of Human Subjects 1979).
44. This link between the three principles and their concrete applications was first discovered by Michael Yesley (see Beauchamp 2005, 17–18).
45. The example of such an attribute can be relationality (see e.g. Hirschmann 1989; see also Levine 2005, 134, note 9).
46. According to commissioner Lebacqz, however, the language adopted by Belmont still leaves a great deal to be desired with respect to the notion of autonomy and the place it occupies in regard to the principle of respect for persons. Although not opting for the strong notion of autonomy in its wording, Belmont, nevertheless allows for and even contributes to the interpretation which reads autonomy in a strong sense. Lebacqz's remarks interestingly counterbalance Churchill's view and supplement mine. See Lebacqz (2005, 100–103).
47. Hereafter I cite Pontifical Council for Pastoral Assistance to Health Care Workers (1995).
48. Hereafter I refer to the text of the *Directives* available on the official website of the U.S. Conference of Catholic Bishops (see National Conference of Catholic Bishops 2001).
49. On the professional-patient relationship as laid out by the *Directives* see Sims (2005).
50. On conscience as related both to individuals and to Catholic institutions see e.g. Brown (2006).
51. A number of helpful remarks on the subject may be found in Beauchamp (2004).
52. Hereafter I quote the text of Flannery (2004, 903–1001).
53. Hereafter I quote the text of the *Catechism* as published in *Catechism of the Catholic Church* (1997).

54. Having said that it nonetheless ought not to be forgotten that in his works following *Basic Christian Ethics* Ramsey demonstrates a greater appreciation of natural law (see e.g. Ramsey 1962, 1968).
55. This transformation of natural law by love is another trait of Ramsey's thought which is vague and hard to defend.
56. Commenting on the covenantal dimension of medical relationship, Ramsey recalls the words of Abraham Lincoln, according to which "no man is good to govern another without his consent." Converted into the research context, these words caution that "no man is good enough to experiment upon another without his consent," an observation analogous to and effective also in a classic, therapeutic physician-patient relationship in which "no man is good enough to cure another without his consent" (Ramsey 2002, 7).
57. For more on this, see for instance Pinckaers (2005b).
58. Accordingly, Ramsey admits that "a choice may be free and responsible despite the fact that it began in emotional bias one way rather than another, and consent can be informed without being encyclopedic" (Ramsey 2002, 3).
59. A useful introduction into the issue of ends of medicine can be found in Pellegrino (2001b).
60. For more on this see Pellegrino and Thomasma (1988, 73–91).
61. Moreover, akin to other natural virtues, prudence does not prove callous to the supernatural; on the contrary, as it operates on the supernatural level, it is given a complementary and supplementary sense, import, and essence. For more on this, see Pellegrino and Thomasma (1996, 17–22, 52, 99–116).
62. After Aristotle, our authors distinguish between licit and illicit self-interest. That investigators "seek to advance their careers, provide for their families, and enjoy the approval of their peers and the gratification of honors, public recognition and leisure" proves legitimate and morally justifiable as long as these factors do not "dominate and interfere with the ends proper to living an ethically good life or achieving the aims of one's work" (Pellegrino and Thomasma 1993, 135).
63. For a synthesis of the investigator's self-interest, its sources and the perils adhering to it, see Pellegrino (2001a, 194–97).
64. This is true especially in the case of *express* consent called in *Principles* "the basic paradigm of autonomy in health care, research, politics and other contexts." Aside from its *express* form, informed consent can also be *tacit, implicit, implied, presumed* and *deferred* (see Beauchamp and Childress 2001, 65–68).
65. "Medical and research codes, as well as case law and federal regulations, have developed models of informed consent that are delineated entirely in a $sense_2$ format, although they have sometimes attempted to justify the rules by appeal to something like $sense_1$" (Faden, Beauchamp, and King 1986, 281).
66. For an extensive analysis of coercion, persuasion, and manipulation, see Faden, Beauchamp, and King (1986, 337–73).
67. First introduced by W.D. Ross and taken over by Beauchamp and Childress, the concept of prima facie duties features a great deal of ambiguity and haziness. In brief, "one has a prima facie duty to do some act just in case there is some reason to think that it is one's duty to do it. A fuller consideration of the act's features might rebut this conclusion by showing that the features providing reason to ascribe the duty are overridden by other features of the act. Ross described each prima facie duty as a 'parti-resultant' attribute, obtained by looking at one morally relevant aspect of an act, whereas 'being one's [actual] duty' is a 'toti-resultant' attribute, obtained by looking together at all of the relevant aspects" (Richardson 2003). For a recent comprehensive critique of principlism, including the notion of prima facie duties see Plich (2006).
68. In line with what has been demonstrated by number of recent studies, while at first glance promising and attractive, *reflective equilibrium,* when examined more closely, proves to be an illusory and misleading method of justification of moral norms. For a thorough analysis thereof, see e.g. Plich (2006, 141–204).

69. For the full list of the virtues critical to biomedical ethics see Beauchamp and Childress (2001, 32–39).
70. On the formation of conscience, see also Grisez (1997a, 82–85).
71. In the U.S. it is agreed that "if the patient's treating physician wishes to treat the patient in an innovative fashion, independent of any research effort, then independent review is not required and informed consent need not be obtained in those cases in which the usual emergency exception to obtaining consent to a therapeutic intervention applies" (Brody 1998, 39). Also, informed consent does not have to be secured if a patient has already given general consent for treatment (see e.g. Truog et al. 1999, 804). This acceptance of innovative therapies in the U.S. harmonizes with the CIOMS Guidelines according to which "a physician may undertake the compassionate use of an investigational therapy before obtaining the approval or clearance of an ethical review committee, provided three criteria are met: a patient needs emergency treatment, there is some evidence of possible effectiveness of the investigational treatment, and there is no other treatment available that is known to be equally effective and superior" (commentary on Guideline 2).
72. In regard to informed consent, both the CIOMS Guideline 4 and the commentary on it use the word "waiver" in the sense of an exception in general. The commentary on Guideline 4, however, provides a "waiver of the requirement of a signed consent form," which pertains to one particular element of the informed-consent process. Still, in Guideline 6 the word "exception" supersedes "waiver" although the meaning and context remain the same. Nevertheless, the commentary on Guideline 6 refers to "exception to the requirement for informed consent in studies of emergency situations" which stands in accord with the wording preferred in the literature on the subject. Interestingly, a similar broad and opaque usage of the word "waiver," disregarding the difference between exception and waiver, can also be found in the Code of Medical Ethics of the American Medical Association (see Council on Ethical and Judicial Affairs 2006, 233–34).
73. Some prominent authors, for example Beauchamp, Childress, Faden, Berg, Lidz, and Appelbaum, follow the U.S. legal regulations and include incompetence as another possible exception from informed consent, but in my estimation this begs the question. In fact, the lack of competence gives grounds for exception in the strict sense when a proxy is not available. It also touches upon therapeutic privilege as the condition likely to be caused by disclosure of information. As is shown in further paragraphs of the current subsection, both incompetence and competence play the role of prerequisites, often decisive ones, in determining and allowing for the three basic forms of exception from informed consent. Therefore, I am convinced that no serious grounds exist to single out one of those prerequisites and upgrade its status to a separate form of consent exception. The proponents of such a status in bioethics literature seem simply to assume it de facto. As an illustration, see for example Faden, Beauchamp, and King (1986, 35); Beauchamp and Childress (2001, 84).
74. Certain bioethicists understand an emergency exception with respect to competent patients more broadly so that it can occur in the situation when a patient could give a valid consent, but in view of his or her condition marked by substantial risk of a serious injury or even death unless an immediate medical action is taken, there is simply no time to secure it (see for example Faden, Beauchamp, and King 1986, 36). Since human dignity of the patient is at stake here, I generally find this interpretation too lenient and indeed one that unnecessarily increases the risk of patient abuses. Even if the implementation of the entire consent procedure is rightly assessed as too time-consuming, usually there is no reason whatsoever why the doctor could not secure at least the oral assent of the patient, or consent following an abbreviated disclosure, before going forward with the procedure. See Berg et al. (2001, 78–79).
75. Other possibilities may include prisoners, substance abusers, carriers of infectious disease, and the mentally ill who may be a danger to others. For more on those categories of patients, their informed refusal, and compulsory treatment see Berg et al. (2001, 227–45).

76. Apart from emergency situations, exceptions may also occur in non-emergency settings, as is the case with the epidemiological research. In studies of this type, various medical and often confidential information on patients recorded by physicians and other health care professionals is utilized. As stated in the CIOMS Guidelines, "for such studies it is usually impracticable to obtain the informed consent of each identifiable patient; an ethical review committee may waive the requirement for informed consent when this is consistent with the requirements of applicable law and provided that there are secure safeguards of confidentiality" (commentary on Guideline 18). For more on consent exception in epidemiological research, see Brody (1998, 55–75). See also Council for International Organizations of Medical Sciences (2005).
77. In contemporary literature on the subject the opinion prevails that the basis for this kind of judgment is the standard of "a reasonable person" rather than the convictions of the particular person involved. In other words, what the presumption encompasses is that since a reasonable person would consent to emergency treatment in given circumstances, any patient would consent to it as well (see Faden, Beauchamp, and King 1986, 36; Berg et al. 2001, 76).
78. Hereafter I refer to the text of the Code available at Code of Federal Regulations (On-line, accessed August 9, 2008).
79. The example offered to illustrate this is the situation in which "a patient refuses to be informed about whether he has a certain genetic disorder" and "there is a clear possibility that he carries the disorder", so that "his refusal could have a deleterious effect on his children" (Brannigan and Boss 2001, 126).
80. Brannigan and Boss, for instance, hold therapeutic privilege appropriate only when "solid evidence exists that disclosure will make the patient so emotionally distraught as to compromise his or her ability to act autonomously" (Brannigan and Boss 2001, 126).
81. For more detailed description of each standard see e.g. Berg et al. (2001, 46–52); Faden, Beauchamp, and King (1986, 30–34, 305–11).
82. The Merriam-Webster Unabridged Dictionary offers the following definition: "a legal instrument authorizing one to act as the attorney or agent of the grantor either generally for the management of a specified business or enterprise or more often specifically for the accomplishment of a particular transaction" ("Power of Attorney" 2002). Interestingly, both definitions suggest documents rather than agents, whereas professional literature, when using the term, refers to either documents or agents, often without a clear distinction between the two.
83. For more on different methods of selecting a surrogate see Brock (2004, 2484–86).
84. To learn more on the issues inherent to advance directives see O'Rourke (2000, 88–91); Finnis (1994).
85. For more on this and other issues concerning honoring wishes, values, and goals, formerly expressed by a now-incompetent patient see Baergen (1995).
86. Referring to the ward's quality of life, the *Saikewicz* court had this to say: "To the extent that this formulation equates the value of life with any measure of the quality of life, we firmly reject it... Rather than reading the judge's formulation in a manner that demeans the value of the life of one who is mentally retarded, the vague, and perhaps ill-chosen, term 'quality of life' should be understood as a reference to the continuing state of pain and disorientation precipitated by the chemotherapy treatment" (Jonsen, Veatch, and Walters 1998, 157).
87. In general, U.S. courts have established three thresholds of evidence: "preponderance of the evidence (greater than 50 percent probability), clear and convincing evidence (greater than 75 percent probability), and beyond a reasonable doubt (greater than 95 –99 percent probability). The latter is reserved for criminal cases and thought to be inappropriate for this setting" (Berg et al. 2001, 116).
88. This does not change the fact, however, that some ethicists simply follow legal solutions without elaborating on ethical basis for SJS. As an example see Brannigan and Boss (2001, 486–89).

89. In the context of securing informed consent from research subjects, Veatch notes that employing the reasonable-person standard implies that "the investigator must tell the subject what he or she would reasonably want to know in order to consent." Consequently, the content of the standard, as Veatch sees it, fluctuates from one investigator to another depending on their desires and wants, or at best on a subjective or personal perception of what counts as reasonable (Veatch 1996a, 48).
90. Elaborating on this broad agreement in his earlier work, *A Theory of Medical Ethics,* Veatch writes the following: "The family is an important institution in all societies. Though its role differs substantially in different cultures, all give families a broad range of decision-making choices. The same principle of autonomy that gives the individual freedom to deviate from the social consensus of what is best may also provide some limited range of familial discretion. For example, parents are given discretion in selecting a school system for their children. They are even permitted to choose alternative systems that most people would not consider to be fully in the best interests of the ward... There are limits, however. The parents cannot choose to have no schooling at all. They cannot refuse needed blood transfusions or other care when the best available judgment is that such care is clearly in the interests of the ward" (Veatch 1981, 211).
91. Buchanan and Brock say that advance directives do not have merely evidentiary value but are also performative. This is a realistic recognition of limitations inherent in advance directives, especially their instructional variant.
92. For more on the problem of assessing well-being from the legal perspective, see Cantor (2005, 160–68).
93. For a critique of Beauchamp's and Childress's prima facie principles see e.g. Wildes (2000, 55–85).
94. In the second (1996) edition of this book, Engelhardt modifies his views somewhat, yet the principle of permission that he now puts forward and sees as overarching does not deviate perceptibly from the principle of autonomy endorsed in the first edition. I owe this observation to Veatch (2003, 182). It was only after his conversion to Orthodox Christianity that Engelhardt's position on numerous matters, including that of autonomy, diametrically changed. For more on this see Engelhardt (2000).
95. In Veatch's own words, the balancing strategy he resorts to, "is not a very satisfying one" as "it does not provide a precise measuring technique permitting the balancing of counterclaims" (Veatch 1981, 304).
96. For more on this, see Veatch (1996a, 47–51). See also Veatch (1996b, 42). This is an interesting claim given current trends in bioethics that seem to acknowledge that if autonomy can ever be overturned, this might be done only by beneficence. See e.g. Jonsen (1996); Ackerman (1996).
97. For a critique thereof, see, e.g. Pellegrino and Thomasma (1988, 12–22, 43–50).
98. This confusing intermingling of objectivity and subjectivity can be also found in Buchanan and Brock when they say, for example, that "self-determination is one of the most plausible candidates for an objective ideal for the person within ideal theory of the good for persons" (Buchanan and Brock 1989, 39).
99. Competence has been given a great deal of attention in today's bioethics. In his *Encyclopedia of Bioethics* entry on competence, Robert Wettstein notes that "individuals have various specific abilities or capacities as well as incapacities each along a continuum. A person is considered incapacitated when the person is no longer able to perform that specific function and incompetent when a court has so ruled. Legally, there is a presumption of competence, which may be overcome when the court is presented with adequate evidence of incapacitation" (Wettstein 2004, 489). It follows that while capacity is determined by medical professionals, the competence remains within the courts' prerogatives. Berg et al. explain that sensu stricto two categories of incompetent persons exist: minors and adults adjudicated incompetent by the court. If the second happens to be the case, there are still various levels at which incompetence can be defined. For instance, the individual can be adjudged

totally incompetent which results in "total legal disability," entitling the guardian to decide in literally all legally relevant matters on the ward's behalf. But incompetence can also be adjudicated at a more specific level. Instead of being rendered incompetent for all purposes, the person in question may retain legal competence over certain aspects of his or her life, while being incompetent in regard to one or a more of them. What is more, "competence to make medical decisions is now almost always considered distinct from competence to perform other tasks, and thus is dealt with through adjudication of specific incompetence" (Berg et al. 2001, 96). In keeping with these observations, Beauchamp and Childress accentuate the specific character of competence which they identify with an ability to carry out a task or a gamut of tasks. Consequently, "*the criteria* of particular competencies vary from context to context because the criteria are relative to specific tasks" (Beauchamp and Childress 2001, 70). What further complicates the matter is that often the level of competence regarding the same specific task fluctuates over the course of time. Depending on the patient's condition it may change considerably from week to week, day to day, or even from hour to hour. As a result, instead of being grasped as a static or inert concept, competence must be looked upon as a dynamic continuum of ability which "runs from full mastery through various levels of partial proficiency to complete ineptitude" (Beauchamp and Childress 2001, 72).

100. From this it follows that "an autonomous person is (necessarily) a competent person (for making decisions) and that judgments about whether a person is competent to authorize or refuse an intervention should be based on whether that person can choose autonomously in particular circumstances" (Beauchamp and Childress 2001, 72).
101. Hereafter I quote the text of *Summa Theologica* after Aquinas (1920).
102. Hereafter all the biblical quotations are taken from *Holy Bible* (2006).
103. A very similar formulation of the Golden Rule was recently used by Pope Benedict XVI: "Do to others as you would have them do to you, and avoid doing what you would not want them to do" (Benedict XVI 2008c).
104. Grisez understands modes of responsibility as intermediate principles "which stand midway between the first principle and the completely specific norms which direct choices" (Grisez 1997a, 189). The fifth mode has the following definition: "One should not, in response to different feelings toward different persons, willingly proceed with a preference for anyone unless the preference is required by intelligible goods themselves" (Grisez 1997a, 211). The first principle, in turn, has been thus explained: "In voluntarily acting for human goods and avoiding what is opposed to them, one ought to choose and otherwise will those and only those possibilities whose willing is compatible with a will toward integral human fulfillment" (Grisez 1997a, 184).
105. For more on this, see e.g., Grisez (1993, 282–84).
106. For a fine analysis of mercy, including its relation to justice and love, see John Paul II (1980b). Upon referring to "the fundamental link between mercy and justice spoken of by the whole biblical tradition, and above all by the messianic mission of Jesus Christ," the Pope goes on to say that "true mercy is, so to speak, the most profound source of justice" (n. 14). What is more, "love, so to speak, conditions justice and, in the final analysis, justice serves love. The primacy and superiority of love vis-a-vis justice—this is a mark of the whole of revelation—are revealed precisely through mercy" (n. 4).
107. Hereafter I quote the text of the document available on the Vatican website (see Benedict XVI 2005).
108. Due to space limitations I cannot indulge in a more detailed analysis of the profound reality of human love. The literature on the topic is vast and such analysis could well be a topic of a separate study. In addition to Pope Benedict XVI's encyclical letter quoted above, a few representatives of research on love are the following: Wojtyla (1981); Melina and Anderson (2006); Balthasar (2004); Rousselot (2001).
109. According to Friedland, in research, "bias is a process during the design, implementation, or analysis of the study by which researchers, subjects or instruments effect a spurious increase or decrease in the magnitude of the finding" (Friedland 1998, 161).

110. "In a longitudinal study the researcher follows a cohort of subjects over time, performing repeated measurements at prescribed intervals. Because the same individuals are tested throughout the study, personal characteristics remain relatively constant, and differences observed over time can be interpreted as developmental change" (Portney and Watkins 2000, 269).
111. "In a cross-sectional study, the researcher studies a stratified group of subjects at one point in time and draws conclusions about development within a population by comparing the characteristics of those strata" (Portney and Watkins 2000, 269).
112. In an example of a classic cohort study given on the same page, "three groups of subjects *without lung cancer* are identified and recruited on the basis of their exposure status, i.e. *current smokers, ex-smokers, and lifetime non-smokers.* These groups would be then followed up (say) for ten years and rates at which they develop lung cancer compared."
113. Prospective research is one in which, "variables are measured through direct recording in the present. The researcher follows patients as they progress through their treatment or evaluation" (Portney and Watkins 2000, 266).
114. As explained by Portney and Watkins, in a case-control study, "groups of individuals are selected on the basis of whether they have the disorder under study. *Cases* are those classified as having the disorder, and *controls* are chosen as a comparison group without the disorder" (Portney and Watkins 2000, 324).
115. "Retrospective research involves the examination of data that have been collected in the past, often obtained from medical records or surveys" (Portney and Watkins 2000, 266).
116. Clinical trials can also be used for testing medical devices. For more on this see Piantadosi (2005, 73–75).
117. Exceptions to this rule include Phase-IV trials that do not contain controls; these will be examined below.
118. The taxonomy breaking down five types of control groups was employed by the U.S. Department of Health and Human Services and is cited in Kopelman (2004b, 2335).
119. For more on randomization and its types, procedures, and mechanics see e.g. Friedman, Furberg, and DeMets (1998, 62–78).
120. This, however, does not change the fact that certain trials, e.g., "those involving most surgical procedures" can be carried out only as unblinded, that is, open. "In an unblinded or open trial, both the participant and investigator know to which intervention the participant has been assigned" (Friedman, Furberg, and DeMets 1998, 82).
121. Double-blinding may be achieved, for example, "by using identical coded capsules, the identity of which is revealed only after completion of the trial" (Friedland 1998, 157).
122. For more on triple-blind trials see Friedman, Furberg, and DeMets (1998, 86).
123. As Friedman et al. point out "this dose is usually referred to as the maximally tolerated dose, or MTD. Much of the literature has discussed how to extrapolate animal model data to the starting dose in humans or how to step-up the dose levels to achieve the MTD" (Friedman, Furberg, and DeMets 1998, 4).
124. As the CIOMS Guidelines state, "most individuals with mental or behavioral disorders are capable of giving informed consent" (Guideline 15).
125. For more on this, see Sunderland and Dukoff (1999, 108–109).
126. For more on this obligation in the context of schizophrenia as well as on the concept of tripartite consent implying supervision and evaluation of the quality of a given informed consent by three independent professionals, see Posever and Chelmow (2001).
127. To learn more on depression, its different types and meanings, see e.g. Thompson (1989, 87–115).
128. For more on this, see Michels (2003, 236).
129. For more on the specifics of using placebos in psychiatric research, see e.g. DuVal (2004, 56–57); Kim (2003); Rothman and Michels (1994).
130. On students as research subjects, see e.g. Pritchard and Koski (2004, 2469–74); Gamble (2003, 266–70); Miller and Kreiner (2008, 53–64).
131. On research involving military personnel, see e.g. Lederer (2004); Moreno (2003).

132. For a concise presentation of research involving prisoners as subjects in twentieth-century United States, see Hornblum (1997).
133. For one significant illustration of this line of argumentation, see Levine (1982).
134. For a detailed list of the elements of coercion in a prison environment, see e.g. Cohen (2003, 262). Cohen also offers a critique of the position that due to coercion free informed consent cannot be secured from incarcerated individuals.
135. For more on the concept of comparative justice, see Montague (1980).
136. Subpart B provides additional safeguards for pregnant women, fetuses, and neonates, whereas subpart D addresses children as research subjects.
137. For more on reasons ethically justifying children serving as research subjects, see e.g. Brock (1994, 86–93).
138. To learn more on the intricacies of children's sharing in the research decision-making, see e.g. Weithorn and Scherer (1994).
139. Further, *Nature* covered not long ago the case of a British boy who was genetically two people because "he was formed when two eggs, fertilized by two different sperm, fused into one embryo inside his mother's womb," providing an example of chimera. Although a hermaphrodite, the chimera-boy enjoyed good health and despite featuring tissues generated within two different embryos, he was one unified, organized body and thus one human subject (see Pearson 2002).
140. For polemical purposes in ethical disputes the pseudo-scientific term "pre-embryo" has been coined and broadly used. See e.g. McCormick (1991); Ashley and Moraczewski (2001, 189).
141. One example of such research is human embryonic stem cell research. See e.g. President's Council on Bioethics (2004); Smith (2007); O'Rourke (2004); Doerflinger (2001).
142. George and Tollefsen propose to modify this formulation because "identical twins do not come into existence at fertilization. At least one twin comes to be at a later point when the embryo divides." For this reason, it appears more appropriate and in accord with scientific facts to say that "the vast majority of us" begin existence at fertilization (George and Tollefsen 2008, 55).
143. For more on this, see e.g. Morowitz and Trefil (1992, 157–59).
144. In regard to this, O'Rourke asks somewhat rhetorically: "but are we to say that historically, when more than half of the infants born died during childbirth, they were never living human beings?" (O'Rourke 2006, 248). This question can also be used to indicate a highly flawed nature of another line of reasoning characteristic of the fourth argument against the human status of human embryo. In brief the point is that "since embryonic death in early miscarriages happens with predictable frequency... we are warranted in concluding that the deliberate destruction of human beings in the embryonic stage is morally acceptable" (George and Tollefsen 2008, 138).
145. In fact, even if the humanity of the early embryo is admitted, its personhood may still be questioned, as certain authors propose that not all human beings are persons. For more on this view and its lack of soundness, see O'Rourke's article on personhood referenced above (O'Rourke 2006). See also Lee and George (2008, 118–39); Lee (1996, 58–62); Grisez (1989).
146. See also Pontifical Academy for Life (2000). This document describes embryonic stem cells and their capacities.
147. Pluripotent stem cells coming from extra-embryonic sources are often called "adult stem cells" to emphasize that their procurement radically differs in terms of moral justifiability from obtaining embryonic stem cells which always presupposes killing the embryo. Commenting upon the data available from the National Institute of Health and the National Marrow Donor Program, May notes that currently "adult stem cells are in established or experimental clinical use to treat human patients with several dozen conditions, including spinal cord injury, multiple sclerosis, cardiac problems, and stroke. For example, paralyzed human persons are being successfully treated with their own adult stem cells" (May 2000, 231). On the following six pages, May elaborates at greater length upon the topic of new

sources for embryonic stem cells, including controversial ANT-OAR project, as well as provides relevant bibliographical references for further research.

148. An excellent source of information on this subject is the website of DoNoHarm: The Coalition of Americans for Research Ethics (On-line, accessed September 4, 2008).
149. O'Rourke expressed the matter well when he pointed out that "the proposed benefits which would result from stem-cell research are not proposed as potential happenings, but rather as sure things that will be accomplished if enough research and federal funding are devoted to the project. In spite of the fact that many of the predicted outcomes are based upon research with mice, spokespersons for stem-cell research, especially research with stem cells taken from embryos, speak as though the golden age of health and well being will be imminent as a result of this type of research" (O'Rourke 2004, 295).
150. For more technical details on how gene alteration works and how it can yield therapeutic effects, see May (2000, 238–40); Prentice (2001, 529–32).
151. For an opposite view, strongly favoring genetic enhancement, see e.g. Green (2007).
152. See "Communion and Stewardship," the document issued by International Theological Commission in 2004, especially n. 91. Hereafter I refer to the text of the document after International Theological Commission (2004).
153. To learn more on how prenatal diagnosis works and what genetic diseases can be tested with it, see Byrnes (2001, 557–59). See also Ashley, DeBlois, and O'Rourke (2006, 98–100); Maher (2001, 519–26).
154. Hereafter I quote the text of this encyclical letter available on the Vatican website (see John Paul II 1995).
155. Hereafter I refer to the text of *Donum Vitae* available on the Vatican website (see Congregation for the Doctrine of the Faith 1987).
156. Both papal and *Donum Vitae*'s teaching has been reiterated by both the *Charter for Health Care Workers* (n. 59–61) and the *Ethical and Religious Directives for Catholic Health Care Services* (n. 50).
157. While in prenatal screening a commonly advanced "solution" to the deficiencies detected in the fetus is abortion, in pre-implantation screening "the affected embryo is discarded, and another, unaffected, embryo is chosen for implantation in the mother's womb" (Byrnes 2001, 560).
158. As indicated by May, the second common method of procuring embryos for screening consists in flushing them out of the uterus by uterine lavage with the intention to eliminate those who may carry serious diseases (May 2000, 244–45).
159. For more on this, see, e.g., Congregation for the Doctrine of Faith (1987, part II, n. 5).
160. "Genetic counseling... is concerned not only with informing the patient and family but also with providing psychologically oriented counseling to help individuals adapt and adjust to the impact and implications of the disorder in the family" (Nussbaum et al. 2004, 375).
161. Hereafter I quote the text of this encyclical letter available on the Vatican website (see Paul VI 1968).
162. Originally this chapter appeared as an article in *Perspectives in Biology and Medicine* 18 (1974, 2–20).
163. Originally this chapter, being the second McCormick's reply to Ramsey, was published in *Hastings Center Report* 6, no. 6 (1976), 41–46).
164. One example of a remote connection to treatment offered by Ramsey refers to an immediate threat of epidemics wherein a healthy child and member of the endangered population may be justifiably enrolled in research aimed at providing remedies against it. No doubt, such research is relevant to the child's situation that calls for protection. In their pursuit thereof, parents or guardians offering consent weigh the threats involved in the trial against the risks inherent to the epidemics disease that imperils the child. Ramsey expresses this as follows: "no one ought to consent for a child to be made the subject of medical investigations primarily for the accumulation of scientific knowledge, except in the face of epidemic conditions that bring upon the individual child proportionately the same or likely greater dangers" (Ramsey 2002, 25).

165. May gives the following reference: Simon, Powers, and Gunnemann (1972, 22–25).
166. Hereafter all the citations of the Commission's recommendations come from National Commission for the Protection of Human Subjects (1977a). The preliminary requirements for ethical acceptability of research on children are, according to the Commission, the following: "(A) the research is scientifically sound and significant; (B) where appropriate, studies have been conducted first on animals and adult humans, then on older children, prior to involving infants; (C) risks are minimized by using the safest procedures consistent with sound research design and by using procedures performed for diagnostic or treatment purposes whenever feasible; (D) adequate provisions are made to protect the privacy of children and their parents, and to maintain confidentiality of data; (E) subjects will be selected in equitable manner" (Recommendation 2).
167. Minimal risk, as defined by the Commission, "is the probability and magnitude of physical or psychological harm that is normally encountered in the daily lives or in the routine medical or psychological examination of healthy children" (National Commission for the Protection of Human Subjects 1977a).
168. In addition, Recommendation 7 charges the IRBs with the duty to ascertain that: (1) the assent of the child was given whenever doable; (2) the permission of parents or guardians was secured, as well; (3) the process of obtaining both (1) and (2) was closely overseen and "at least one parent or guardian was involved in the conduct of research." Additionally, "a child's objection to participation in research should be binding unless the intervention holds out a prospect of direct benefit that is important to the health or well-being of the child and is available only in the context of research." Recommendation 8, in turn, provides further for "conditions or a subject population for which parental or guardian permission is not a reasonable requirement to protect the subjects" and therefore can be waived. The example might be research related to some conditions in teenagers such as drug abuse, pregnancy, or venereal diseases.
169. See the dissenting statements of Commissioners Cooke and Turtle (National Commission for the Protection of Human Subjects 1977a, 145–53).
170. For more on this, see Bartholome (1977a).
171. Friedman-Ross notes that according to Jean Lockhart, "this phrase was coined by H. Shirkey in 1963... and it refers to those persons with rare conditions that do not receive adequate clinical study and understanding. The result is that clinicians treat these subject-patients using unproven therapies. So, for example, if new medicines could not be tested on children, then pediatricians would have to rely on adult data which may or may not be appropriate for children, who metabolize drugs differently" (Friedman Ross 1998, 103, note 14; see also Lockhart 1977).
172. May borrows the principle of fairness from Grisez and incorporates it into his own analyses. In Grisez's account, as discussed earlier in this study, fairness plays a critical role in the application of the Golden Rule, as it underlies impartiality and belongs among the most important facets of justice.
173. Quotation from Benedict M. Ashley's e-mail message to author, July 30, 2008.
174. For an extended and profound analysis of the relationship between faith and reason, see John Paul II (1998).
175. This passage is also quoted in the *Catechism of the Catholic Church*, n. 159.
176. As explained in the note immediately preceding the Principles and Guidelines' contents, the document "was produced by The Joint Task Force on Ethics in Clinical Research, comprising members of the Catholic Medical Association and the ethicists of the National Catholic Bioethics Center" with its aim being "to aid clinical researchers in ethical decision making during the conduct of clinical research." Throughout this chapter I quote the text of the Principles and Guidelines as originally published (The Catholic Principles and Guidelines for Clinical Research 2007).
177. This text was a keynote address of the Tenth Bishops' Workshop of the National Catholic Bioethics Center on *Catholic Conscience: Foundation and Formation*, February 1991.

178. Both natural and supernatural virtues have their share in reaching this goal. For more on how Christian virtues affect the principles, see Pellegrino and Thomasma (1996, 23–25, 117–38).
179. As noted in Chapter 3, Aquinas explicitly says this in ST, I-II, q. 94, a. 3.
180. For a comprehensive critique of the arguments rejecting moral absolutes, including that of McCormick with further bibliography, see e.g. May (2003a, 141–83). See also Kaczor (2002); Lee (1981).
181. For more on this, see e.g. Pinckaers (2005a, 156–61).
182. This passage is quoted in the *Catechism of the Catholic Church*, n. 1906.
183. This, I think, should also be seen in the light of the following passage from *Gaudium et Spes*: "the norm of human activity is this: that in accord with the divine plan and will, it harmonize with the genuine good of the human race, and that it allow men as individuals and as members of society to pursue their total vocation and fulfill it" (n. 35).
184. In relation to this some Thomists, such as Charles De Koninck, argue for the supremacy of the common good over the private good (see e.g. De Koninck 1943, 9–109; De Koninck 1945).

References

Ackerman, T. F. 1996. "Choosing Between Nuremberg and the National Commission: The Balancing of Moral Principles in Clinical Research." In *The Ethics of Research Involving Human Subjects: Facing the 21st Century*, edited by H. Y. Vanderpool, 83–104. Frederick, MD: University Pub. Group.

Ainslie, D. C. 2004. "Principlism." In *Encyclopedia of Bioethics*, edited by S. G. Post, 2099–104. New York, NY: Macmillan Reference USA.

Ajlouni, K. M. 1995. "History of Informed Medical Consent." *The Lancet* 346:980.

Alexander, L. 1976. "Ethics of Human Experimentation." *Psychiatric Journal of the University of Ottawa* 1:40–46.

Angell, M. 1997. "The Ethics of Clinical Research in the Third World." *The New England Journal of Medicine* 337(12):847–49.

Angell, M. 2000. "Investigators' Responsibilities for Human Subjects in Developing Countries." *The New England Journal of Medicine* 342(13):967–69.

Annas, G. J. 1992. "The Nuremberg Code in U.S. Courts: Ethics Versus Expediency." In *The Nazi Doctors and the Nuremberg Code: Human Rights in Human Experimentation*, edited by G. J. Annas and M. A. Grodin, 201–22. New York, NY: Oxford University Press.

Annas, G. J., L. H. Glantz, and B. F. Katz. 1977. *Informed Consent to Human Experimentation: The Subject's Dilemma*. Cambridge, MA: Ballinger.

Annas, G. J., and M. A. Grodin, eds. 1992. *The Nazi Doctors and the Nuremberg Code: Human Rights in Human Experimentation*. New York, NY: Oxford University Press.

Annas, G. J., and M. A. Grodin. 1998. "Human Rights and Maternal-Fetal HIV Transmission Prevention Trials in Africa." *American Journal of Public Health* 88(4):560–63.

Aquinas, T. 1920. *Summa Theologica*. Eng. trans. Fathers of the English Dominican Province, London: Benzinger Brothers.

Aristotle. 1944. *Politics*. Eng. trans. H. Rackham. Cambridge, MA: Harvard University Press.

Aristotle. 1962. *The Nicomachean Ethics*. Eng. trans. H. Rackham. Cambridge, MA: Harvard University Press.

Ashley, B. M., J. DeBlois, and K. D. O'Rourke. 2006. *Health Care Ethics: A Catholic Theological Analysis* (5th ed.). Washington, DC: Georgetown University Press.

Ashley, B. M., and A. Moraczewski. 2001. "Cloning, Aquinas, and the Embryonic Person." *The National Catholic Bioethics Quarterly* 1(2):189–201.

Ashley, B. M., and K. D. O'Rourke. 1997. *Health Care Ethics: A Theological Analysis* (4th ed.). Washington, DC: Georgetown University Press.

Ashley, B. M., and K. D. O'Rourke. 2002. *Ethics of Health Care: An Introductory Textbook* (3rd ed.). Washington, DC: Georgetown University Press.

"Attorney, Power Of". 2008. In *Encyclopaedia Britannica Ultimate Reference Suite Edition*.

Baergen, R. 1995. "Revising the Substituted Judgment Standard." *The Journal of Clinical Ethics* 6(1):30–38.

Balthasar, H. U. von. 2004. *Love Alone Is Credible*. San Francisco, CA: Ignatius Press.

Barrett, B., D. Muller, D. Rakel, D. Rabago, L. Marchand, and J. Scheder. 2006. "Placebo, Meaning, and Health." *Perspectives in Biology and Medicine* 49(2):178–99.

Bartholome, W. G. 1977a. "Proxy Consent in the Medical Context: The Infant as Person." In National Commission for the Protection of Human Subjects of Biomedical and Behavioral Research, *Appendix: Research Involving Children. DHEW Publication No. (OS) 77-0005*, sec. 3, 23–24. Washington, DC: US Government Printing Office.

Bartholome, W. G. 1977b. "The Ethics of Non-Therapeutic Clinical Research on Children." In National Commission for the Protection of Human Subjects of Biomedical and Behavioral Research, *Appendix: Research Involving Children. DHEW Publication No. (OS) 77-0005*, sec. 3, 1–19. Washington DC: US Government Printing Office.

Baumslag, N. 2005. *Murderous Medicine: Nazi Doctors, Human Experimentation, and Typhus*. Westport, CT: Praeger Publishers.

Beauchamp, T. L. 2004. "Paternalism." In *Encyclopedia of Bioethics*, edited by S. G. Post, 1983–90. New York, NY: Macmillan Reference USA.

Beauchamp, T. L. 2005. "The Origins and Evolution of the Belmont Report." In *Belmont Revisited: Ethical Principles for Research with Human Subjects*, edited by J. F. Childress, E. M. Meslin, and H. T. Shapiro, 12–25. Washington, DC: Georgetown University Press.

Beauchamp, T. L., and J. F. Childress. 2001. *Principles of Biomedical Ethics* (5th ed.). New York, NY: Oxford University Press.

Beauchamp, T. L., and R. R. Faden. 2004. "Informed Consent. History." In *Encyclopedia of Bioethics*, edited by S. G. Post, 1271–76. New York, NY: Macmillan Reference USA.

"Beaumont, William". 2008. In *Encyclopaedia Britannica Ultimate Reference Suite Edition.*

Beecher, H. 1960. "Human Experimentations – A World Problem from the Standpoint of a Medical Investigator." *World Medical Journal* 8:79–80.

Beecher, H. 1966. "Ethics and Clinical Research." *The New England Journal of Medicine* 274(24):1351–60.

Beecher, H. 1970. *Research and the Individual*. Boston, MA: Little, Brown & Co.

Benedict XVI. 2005. *Encyclical Letter "Deus Caritas Est"* [On-line] Accessed August 29, 2008. http://www.vatican.va/holy_father/benedict_xvi/encyclicals/documents/hf_ben-xvi_enc_20051225_deus-caritas-est_en.html

Benedict XVI. 2008a. *Address to Catholic Educators at the Catholic University of America* [On-line] Accessed August 16, 2008. http://www.vatican.va/holy_father/benedict_xvi/speeches/2008/april/documents/hf_ben-xvi_spe_20080417_cath-univ-washington_en.html

Benedict XVI. 2008b. *Address to the Members of the General Assembly of the United Nations Organization* [On-line] Accessed August 19, 2008. http://www.vatican.va/holy_father/benedict_xvi/speeches/2008/april/documents/hf_ben-xvi_spe_20080418_un-visit_en.html

Benedict XVI. 2008c. *Video-Message to Catholic and People of the United States of America on the Occasion of the Upcoming Apostolic Journey* [On-line] Accessed April 17, 2008. http://www.vatican.va/holy_father/benedict_xvi/messages/pont-messages/2008/documents/hf_ben-xvi_mes_20080408_people-usa_en.html

Berg, J. W., P. S. Appelbaum, Ch. W. Lidz, and L. S. Parker. 2001. *Informed Consent: Legal Theory and Clinical Practice* (2nd ed.). Oxford and New York, NY: Oxford University Press.

Bernard, C. 1927. *Introduction to the Study of Experimental Medicine*. Eng. trans. H. Copley. New York, NY: Macmillan.

Bhutta, Z. A. 2004. "Beyond Informed Consent." *Bulletin of the World Health Organization* 82(10):771–77.

Bompiani, A. 2003. "Experimental Research in the Field of Biomedicine. Spheres, Methods, Validity Criteria for Research Projects." In *Ethics of Biomedical Research in a Christian Vision. Proceedings of the Ninth Assembly of the Pontifical Academy for Life (24–26 February, 2003)*, edited by J. De Dios Vial Correa and E. Sgreccia, 93–125. Vatican City: Libreria Editrice Vaticana.

Bourke, V. J. 2008. *History of Ethics*, vol. 1. Mount Jackson, VA: Axios Press.

Brannigan, M. C., and J. A. Boss. 2001. *Healthcare Ethics in a Diverse Society*. Mountain View, CA: Mayfield Pub. Co.

Branson, R. 2004. "Prisoners as Research Subjects." In *Encyclopedia of Bioethics*, edited by S. G. Post, 2104–108. New York, NY: Macmillan Reference USA.

Brennan, T. A. 1999. "Proposed Revisions to the Declaration of Helsinki – Will They Weaken the Ethical Principles Underlying Human Research?." *The New England Journal of Medicine* 341(7):527–31.

Brock, D. W. 1994. "Ethical Issues in Exposing Children to Risks in Research." In *Children as Research Subjects*, edited by M. A. Grodin and L. H. Glantz, 81–101. New York, NY: Oxford University Press, Inc.

Brock, D. W. 2004. "Surrogate Decision-Making." In *Encyclopedia of Bioethics*, edited by S. G. Post, 2483–86. New York, NY: Macmillan Reference USA.

Brody, B. A. 1998. *The Ethics of Biomedical Research: An International Perspective*. New York, NY: Oxford University Press.

Brody, H. 2004. "Placebo." In *Encyclopedia of Bioethics*, edited by S. G. Post, 2030–32. New York, NY: Macmillan Reference USA.

Brown, G. T. 2006. "Conscience of a Catholic Institution: Growing in Catholic Identity." *Ethics & Medics* 31(8):3–4.

Buchanan, A. E., and D. W. Brock. 1989. *Deciding for Others: The Ethics of Surrogate Decisionmaking*. Cambridge and New York, NY: Cambridge University Press.

Byrnes, W. M. 2001. "Human Genetic Technology, Eugenics, and Social Justice." *The National Catholic Bioethics Quarterly* 1(4):555–81.

Cantor, N. L. 2005. "The Bane of Surrogate Decision-Making. Defining the Best Interests of Never-Competent Persons." *The Journal of Legal Medicine* 26:155–205.

Catechism of the Catholic Church (2nd ed.). 1997. New York, NY: Doubleday.

Charland, L. 2008. "Decision-Making Capacity." In *Stanford Encyclopedia of Philosophy* [On-line] Accessed March 4, 2008. http://plato.stanford.edu/entries/decision-capacity

Childress, J. F. 1990. "The Place of Autonomy in Bioethics." *The Hastings Center Report* 20(1):12–17.

Churchill, L. R. 2005. "Toward a More Robust Autonomy. Revisiting the Belmont Report." In *Belmont Revisited: Ethical Principles for Research with Human Subjects*, edited by J. F. Childress, E. M. Meslin, and H. T. Shapiro, 111–25. Washington, DC: Georgetown University Press.

Cochran, C. E. 2007. "Genetic Engineering." In *Encyclopedia of Catholic Social Thought, Social Science, and Social Policy*, vol. 1, edited by M. L. Coulter, S. M. Krason, R. S. Myers, and J. A. Varacalli, 465–66. Lanham, MD: The Scarecrow Press, Inc.

"Code of Federal Regulations" [On-line] Accessed August 9, 2008. http://www.gpoaccess.gov/cfr/index.html

Cohen, C. 2003. "Medical Experimentation on Prisoners." In *Ethical and Regulatory Aspects of Clinical Research: Readings and Commentary*, edited by E. J. Emanuel, R. A. Crouch, J. D. Arras, J. D. Moreno, and Ch. Grady, 262–66. Baltimore, MD: Johns Hopkins University Press.

Colombo, R. 2003. "The Most Vulnerable Subjects of Biomedical Research: *The Case of Human Embryo*." In *Ethics of Biomedical Research in a Christian Vision. Proceedings of the Ninth Assembly of the Pontifical Academy for Life (24–26 February, 2003)*, edited by J. De Dios Vial Correa and E. Sgreccia, 208–33. Vatican City: Libreria Editrice Vaticana.

Congregation for the Doctrine of the Faith. 1987. *Instruction on Respect for Human Life in Its Origin and on the Dignity of Procreation. Replies to Certain Questions of the Day* [On-line] Accessed August 1, 2008. http://www.vatican.va/roman_curia/congregations/cfaith/documents/rc_con_cfaith_doc_19870222_respect-for-human-life_en.html

Connor, E. M., R. S. Sperling, R. Gelber, P. Kiselev, G. Scott, M. J. O'Sullivan, R. VanDyke, M. Bey, W. Shearer, R. L. Jacobson, E. Jimenez, E. O'Neill, B. Bazin, J.-F. Delfraissy, M. Culnane, R. Coombs, M. Elkins, J. Moye, P. Stratton, J. Balsley, and AIDS Clinical Trials Group Protocol Study Group The Pediatric. 1994. "Reduction of Maternal-Infant Transmission of Human Immunodeficiency Virus Type 1 with Zidovudine Treatment." *The New England Journal of Medicine* 331(18):1173–80.

Council for International Organizations of Medical Sciences. 1974. *Protection of Human Rights in Light of Scientific and Technological Progress in Biology and Medicine*. Geneva: World Health Organization.

Council for International Organizations of Medical Sciences. 2002. *International Ethical Guidelines for Biomedical Research Involving Human Subjects*. Geneva: CIOMS.

Council for International Organizations of Medical Sciences. 2005. *Management of Safety Information from Clinical Trials*. Geneva: CIOMS.

Council on Ethical and Judicial Affairs of American Medical Association. 2006. *Code of Medical Ethics: Current Opinions with Annotations*. Chicago, IL: AMA Press.

De Koninck, Ch. 1943. *De La Primaute Du Bien Commun Contre Les Personalistes. Le Principe De L'ordre Nouveau*. Quebec: Editions de l'Universite Laval.

De Koninck, Ch. 1945. "In Defense of St. Thomas: A Reply to Fr. Eschmann's Attack on the Primacy of the Common Good." *Laval Theologique et Philosophique* 1:9–109.

"Department of Defense Request for Exemption from Informed Consent Requirements for Operation Desert Shield and Food and Drug Administration Response." 1990. *Federal Register* 55:52813–17.

Di Blasi, F. 2006. *God and the Natural Law: A Rereading of Thomas Aquinas*. Eng. trans. D. Thunder. South Bend, IN: St. Augustine's Press.

Doerflinger, R. 2001. "The Policy and Politics of Embryonic Stem Cell Research." *The National Catholic Bioethics Quarterly* 1(2):135–43.

Doerflinger, R. 2004. "Experimentation on Human Subjects and Stem Cell Research." In *Moral Issues in Catholic Health Care*, edited by K. McMahon, 93–107. Philadelphia, PA: St. Charles Borromeo Seminary.

Donagan, A. 1977. *The Theory of Morality*. Chicago, IL: University of Chicago Press.

DoNoHarm: The Coalition of Americans for Research Ethics [On-line] Accessed September 4, 2008. http://www.stemcellresearch.org

Drinan, R. F. 1992. "The Nuremberg Principles in International Law." In *The Nazi Doctors and the Nuremberg Code: Human Rights in Human Experimentation*, edited by G. J. Annas and M. A. Grodin, 174–82. New York, NY: Oxford University Press.

DuVal, G. 2004. "Ethics in Psychiatric Research: Study Design Issues." *The Canadian Journal of Psychiatry* 49(1):55–59.

Eberl, J. T. 2006. *Thomistic Principles and Bioethics*. London and New York, NY: Routledge.

Elliott, C. 2003. "Are Severely Depressed Patients Competent to Consent to Research?." In *Ethical and Regulatory Aspects of Clinical Research: Readings and Commentary*, edited by E. J. Emanuel, R. A. Crouch, J. D. Arras, J. D. Moreno, and Ch. Grady, 237–40. Baltimore, MD: Johns Hopkins University Press.

Elliott, C., and S. Parry. 2004. "Mentally Disabled and Mentally Ill Persons. Research Issues." In *Encyclopedia of Bioethics*, edited by S. G. Post, 1825–30. New York, NY: Macmillan Reference USA.

Emanuel, E. J., R. A. Crouch, J. D. Arras, J. D. Moreno, and Ch. Grady, eds. 2003. *Ethical and Regulatory Aspects of Clinical Research: Readings and Commentary*. Baltimore, MD: Johns Hopkins University Press.

Engelhardt, H. T. 1979. "Basic Ethical Principles in the Conduct of Biomedical and Behavioral Research Involving Human Subjects." In National Commission for the Protection of Human Subjects of Biomedical and Behavioral Research, *The Belmont Report: Ethical Principles and Guidelines for the Protection of Human Subjects of Research. DHEW Publication No. (OS) 78-0013*, Appendix, vol. 1, sec. 8, 1–45. Washington, DC: US Government Printing Office.

Engelhardt, H. T. 1986. *The Foundations of Bioethics* (1st ed.). New York, NY: Oxford University Press.

Engelhardt, H. T. 1996. *The Foundations of Bioethics* (2nd ed.). New York, NY: Oxford University Press.

Engelhardt, H. T. 2000. *The Foundations of Christian Bioethics*. Lisse, Netherlands and Exton, PA: Swets & Zeitlinger.

European Agency for the Evaluation of Medical Products. 2001. *EMEA/CPMP Position Statement on the Use of Placebo in Clinical Trials with Regard to the Revised Declaration of Helsinki* [On-line] Accessed August 10, 2008. http://www.emea.europa.eu/pdfs/human/press/pos/1742401en.pdf

Faden, R. R., T. L. Beauchamp, and N. M. P. King. 1986. *A History and Theory of Informed Consent*. New York, NY: Oxford University Press.

Ferriman, A. 2001. "WMA Agrees to Refine Changes to Declaration of Helsinki." *British Medical Journal* 322:1142.

Finnis, J. 1994. "'Living Will' Legislation." In *Euthanasia, Clinical Practice and the Law*, edited by L. Gormally, 167–76. London: The Linacre Centre.

Fischer, G. S., J. A. Tulsky, and R. M. Arnold. 2004. "Advance Directives and Advance Care Planning." In *Encyclopedia of Bioethics*, edited by S. G. Post, 74–79. New York, NY: Macmillan Reference USA.

Fisher, A., O.P. 2007. "The Moral Conscience in Ethics and the Contemporary Crisis of Authority." Paper presented at the International Congress *The Christian Conscience in Support of the Right to Life* at the XIII General Assembly of the Pontifical Academy for Life, Vatican City, February 23–25 [On-line] Accessed August 18, 2008. www.academiavita.org/italiano/AssembleaGenerale/2007/rel2007/eng/fishereng2007.pdf

Fitzgerald, D. W., C. Marotte, R. I. Verdier, W. Johnson, and J. W. Pape. 2002. "Comprehension During Informed Consent in a Less-Developed Country." *The Lancet* 360:1301–302.

Flannery, A., ed. 2004. *Vatican Council II. The Conciliar and Post Conciliar Documents*, vol. 1. Dublin and Northport, NY: Dominican Publications & Costello Publishing Company.

Fletcher, J. C. 1983. "The Evolution of the Ethics of Informed Consent." In *Research Ethics*, edited by K. Berg and K. E. Tranoy, 187–228. New York, NY: Alan R. Liss.

Frankena, W. K. 1973. *Ethics* (2nd ed.). Englewood Cliffs, NJ: Prentice-Hall.

Friedland, D. J. 1998. "Guide for Assessing the Validity of a Study." In *Evidence-Based Medicine: A Framework for Clinical Practice*, edited by D. J. Friedland, 151–82. New York, NY: McGraw-Hill.

Friedman, L. M., C. Furberg, and D. L. DeMets. 1998. *Fundamentals of Clinical Trials* (3rd ed.). New York, NY: Springer.

Friedman Ross, L. 1998. *Children, Families, and Health Care Decision Making*. Oxford: Clarendon Press.

Friedman Ross, L. 2006. *Children in Medical Research: Access Versus Protection*. New York, NY: Oxford University Press.

Gamble, H. F. 2003. "Students, Grades and Informed Consent." In *Ethical and Regulatory Aspects of Clinical Research: Readings and Commentary*, edited by E. J. Emanuel, R. A. Crouch, J. D. Arras, J. D. Moreno, and Ch. Grady, 266–70. Baltimore, MD: Johns Hopkins University Press.

George, R. P., and C. Tollefsen. 2008. *Embryo: A Defense of Human Life*. New York, NY: Doubleday.

Glantz, L. H. 1992. "The Influence of the Nuremberg Code on U.S. Statutes and Regulations." In *The Nazi Doctors and the Nuremberg Code: Human Rights in Human Experimentation*, edited by G. J. Annas and M. Grodin, 183–200. New York, NY: Oxford University Press.

Gotzsche, P. C. 1994. "Is There Logic in the Placebo?." *The Lancet* 344(8927):925–27.

Green, R. M. 2007. *Babies by Design: The Ethics of Genetic Choice*. New Haven, CT: Yale University Press.

Griese, O. N. 1987. *Catholic Identity in Health Care: Principles and Practice*. Braintree, MA: Pope John Center.

Grisez, G. 1989. "When Do People Begin?." *Proceedings of the American Catholic Philosophical Association* 63:27–47.

Grisez, G. 1993. *The Way of the Lord Jesus. Living a Christian Life,* vol. 2. Quincy, IL: Franciscan Press.

Grisez, G. 1997a. *The Way of the Lord Jesus. Christian Moral Principles*, vol. 1. Quincy, IL: Franciscan Press.

Grisez, G. 1997b. *The Way of the Lord Jesus. Difficult Moral Questions*, vol. 3. Quincy, IL: Franciscan Press.

Grodin, M. A. 1992. "Historical Origins of the Nuremberg Code." In *The Nazi Doctors and the Nuremberg Code: Human Rights in Human Experimentation*, edited by G. J. Annas and M. Grodin, 121–44. New York, NY: Oxford University Press.

Haakonssen, L. 1997. *Medicine and Morals in the Enlightenment: John Gregory, Thomas Percival and Benjamin Rush*. Amsterdam, Netherlands and Atlanta, GA: Editions Rodopi B.V.

Harris, J. M., D. Morgan, and M. Ford. 2004. "Embryo and Fetus. Stem Cell Research and Therapy." In *Encyclopedia of Bioethics*, edited by S. G. Post, 722–32. New York, NY: Macmillan Reference USA.

Hauerwas, S., and D. S. Long. 1993. "Foreword." In *Basic Christian Ethics*, edited by P. Ramsey, xiii–xxix. Louisville, KY: Westminster/John Knox Press.

Healy, E. F. 1956. *Medical Ethics*. Chicago, IL: Loyola University Press.

Heller, J. 1972. "Syphilis Victims in U.S. Study Went Untreated for 40 Years." *New York Times* July 26, 1, 8.

Herranz, G. 1998. "The Inclusion of the Ten Principles of Nuremberg in Professional Codes of Ethics: An International Comparison." In *Ethics Codes in Medicine: Foundations and Achievements of Codification Since 1947*, edited by U. Tröhler, S. Reiter-Theil, and E. Herych, 127–39. Aldershot and Brookfield, VT: Ashgate.

Herranz, G. 2004. "Some Christians Contributions to the Ethics of Biomedical Research: A Historical Perspective." In *Ethics of Biomedical Research in a Christian Vision: Proceedings of the Ninth Assembly of the Pontifical Academy for Life: Vatican City, 24–26 February 2003*, edited by E. Sgreccia and J. de Dios Val Correa, 126–41. Città del Vaticano: Libreria Editrice Vaticana.

Hirschmann, N. J. 1989. "Freedom, Recognition, and Obligation: A Feminist Approach to Political Theory." *American Political Science Review* 83(4):1227–44.

Hoffman, G. A., A. Harrington, and H. L. Fields 2005. "Pain and the Placebo: What We Have Learned." *Perspectives in Biology and Medicine* 48(2):248–66.

Holy Bible. Revised Standard Version (2nd Catholic ed.). 2006. San Francisco, CA: Ignatius Press.

Holy See. 1983. *Charter of the Rights of the Family* [On-line] Accessed August 1, 2008. http://www.vatican.va/roman_curia/pontifical_councils/family/documents/rc_pc_family_doc_19831022_family-rights_en.html

Honings, B. 1996. "The Charter for Health Care Workers: A Synthesis of Hippocratic Ethics and Christian Morality." *Dolentium Hominum* 31(1):48–52.

Hornblum, A. M. 1997. "They Were Cheap and Available: Prisoners as Research Subjects in Twentieth Century America." *British Medical Journal* 315:1437–40.

Howard-Jones, N. 1982. "Human Experimentation in Historical and Ethical Perspectives." *Social Science and Medicine* 16:1429–48.

Human, D., and S. S. Fluss. 2001. *The World Medical Association's Declaration of Helsinki: Historical and Contemporary Perspectives. Fifth Draft: 24/07/01* [On-line] Accessed August 11–12, 2008. http://www.wma.net/e/ethicsunit/pdf/draft_historical_contemporary_perspectives.pdf

Hyder, A. A., S. A. Wali, A. N. Khan, N. B. Teoh, N. E. Kass, and L. Dawson. 2004. "Ethical Review of Health Research: A Perspective from Developing Country Researchers." *Journal of Medical Ethics* 30(1):68–72.

International Theological Commission. 2004. *Communion and Stewardship: Human Persons Created in the Image of God* [On-line] Accessed August 12, 2008. http://www.vatican.va/roman_curia/congregations/cfaith/cti_documents/rc_con_cfaith_doc_20040723_communion-stewardship_en.html

John Paul II. 1980a. "Address to Representatives of the Italian Society of Medicine and the Italian Society of General Surgery." *Insegnamenti* III/2, no. 5.

John Paul II. 1980b. *Encyclical Letter "Dives in Misericordia"* [On-line] Accessed May 20, 2008. http://www.vatican.va/holy_father/john_paul_ii/encyclicals/documents/hf_jp-ii_enc_30111980_dives-in-misericordia_en.html

John Paul II. 1982a. *Address to a Meeting of the Pontifical Academy of Sciences* [On-line] Accessed August 1, 2008. http://www.vatican.va/holy_father/john_paul_ii/speeches/1982/october/documents/hf_jp-ii_spe_19821023_pont-accademia-scienze_en.html

John Paul II. 1982b. "To the World Congress of Catholic Doctors." *Insegnamenti* V/3, no. 4.

John Paul II. 1985. "To the Participants at Two Congresses on Medicine and Surgery." *Insegnamenti* III/2, no. 5.

John Paul II. 1986. "To a Conference on Pharmacy in the Synod Hall." *Insegnamenti* IX/2.

John Paul II. 1993. *Encyclical Letter "Veritatis Splendor"* [On-line] Accessed August 16, 2008. http://www.vatican.va/holy_father/john_paul_ii/encyclicals/documents/hf_jp-ii_enc_06081993_veritatis-splendor_en.html

John Paul II. 1995. *Encyclical Letter on the Value and Inviolability of Human Life "Evangelium Vitae"* [On-line] Accessed August 6, 2008. http://www.vatican.va/holy_father/john_paul_ii/encyclicals/documents/hf_jp-ii_enc_25031995_evangelium-vitae_en.html

John Paul II. 1996. *Address to Bishops of the Episcopal Conference of Lesotho on Their "Ad Limina" Visit* [On-line] Accessed August 13, 2008. http://www.vatican.va/holy_father/john_paul_ii/speeches/1996/april/documents/hf_jp-ii_spe_19960416_bishops-lesotho_en.html

John Paul II. 1997. *Address During the Meeting with the Rectors of the Polish Universities* [On-line] Accessed August 16, 2008. http://www.vatican.va/holy_father/john_paul_ii/travels/documents/hf_jp-ii_spe_08061997_universities_en.html

John Paul II. 1998. *Encyclical Letter "Fides et Ratio"* [On-line] Accessed August 16, 2008. http://www.vatican.va/holy_father/john_paul_ii/encyclicals/documents/hf_jp-ii_enc_15101998_fides-et-ratio_en.html

John Paul II. 1999. *Message for the World Day of the Sick for the Year 2000* [On-line] Accessed August 20, 2008. http://www.vatican.va/holy_father/john_paul_ii/messages/sick/documents/hf_jp-ii_mes_19990806_world-day-of-the-sick-2000_en.html

John Paul II. 2000. *Address to the 18th International Congress of the Transplantation Society* [On-line] Accessed August 14, 2008. http://www.vatican.va/holy_father/john_paul_ii/speeches/2000/jul-sep/documents/hf_jp-ii_spe_20000829_transplants_en.html

John Paul II. 2004. *Letter on the Occasion of the 23rd National Congress of the Italian Catholic Physicians' Association* [On-line] Accessed August 16, 2008. http://www.vatican.va/holy_father/john_paul_ii/letters/2004/documents/hf_jp-ii_let_20041109_medici-cattolici_en.html

Jonsen, A. R. 1996. "The Weight and Weighing of Ethical Principles." In *The Ethics of Research Involving Human Subjects: Facing the 21st Century*, edited by H. Y. Vanderpool, 59–82. Frederick, MD: University Pub Group.

Jonsen, A. R. 2003. *The Birth of Bioethics*. New York, NY: Oxford University Press.

Jonsen, A. R. 2005. "On the Origins and Future of the Belmont Report." In *Belmont Revisited: Ethical Principles for Research with Human Subjects*, edited by J. F. Childress, E. M. Meslin, and H. T. Shapiro, 3–11. Washington, DC: Georgetown University Press.

Jonsen, A. R., R. M. Veatch, and L. R. Walters, eds. 1998. *Source Book in Bioethics*. Washington, DC: Georgetown University Press.

Kaczor, Ch. 2002. *Proportionalism and the Natural Law Tradition*. Washington, DC: The Catholic University of America Press.

Katz, J. 1972. *Experimentation with Human Beings. The Authority of the Investigator, Subject, Professions, and State in the Human Experimentation Process*. New York, NY: Russell Sage Foundation.

Katz, J. 2002. *The Silent World of Doctor and Patient*. Baltimore, MD: Johns Hopkins University Press.

Kelly, D. F. 2004. *Contemporary Catholic Health Care Ethics*. Washington, DC: Georgetown University Press.

Kelly, G. A. 1958. *Medico-Moral Problems*. St. Louis, MO: Catholic Hospital Association of the United States and Canada.

Kent, D. M., D. M. Mwamburi, M. L. Bennish, B. Kupelnick, and J. P. A. Ioannidis. 2004. "Clinical Trials in Sub-Saharan Africa and Established Standards of Care: A Systematic Review of HIV, Tuberculosis, and Malaria Trials." *The Journal of the American Medical Association* 292(2):237–42.

Kim, S. Y. H. 2003. "Benefits and Burdens of Placebos in Psychiatric Research." *Psychopharmacology* 171:13–18.

Kim, S. Y. H., and P. S. Appelbaum. 2006. "The Capacity to Appoint a Proxy and the Possibility of Concurrent Proxy Directives." *Behavioral Science & the Law* 24:469–78.

Kimura, R. 1998. "Crimes Against Humanity: The Forgotten Story of Japan." In *Ethics Codes in Medicine: Foundations and Achievements of Codification Since 1947*, edited by U. Tröhler, S. Reiter-Theil, and E. Herych, 119–26. Aldershot and Brookfield, VT: Ashgate Pub.

King, P. A. 2005. "Justice Beyond Belmont." In *Belmont Revisited: Ethical Principles for Research with Human Subjects*, edited by J. F. Childress, E. M. Meslin, and H. T. Shapiro, 136–47. Washington, DC: Georgetown University Press.

Kopelman, L. M. 2004a. "Children. Healthcare and Research Issues." In *Encyclopedia of Bioethics*, edited by S. G. Post, 387–98. New York, NY: Macmillan Reference USA.

Kopelman, L. M. 2004b. "Research Methodology. Clinical Trials." In *Encyclopedia of Bioethics*, edited by S. G. Post, 2334–43. New York, NY: Macmillan Reference USA.

Lebacqz, K. 2005. "We Sure Are Older but Are We Wiser?." In *Belmont Revisited: Ethical Principles for Research with Human Subjects*, edited by J. F. Childress, E. M. Meslin, and H. T. Shapiro, 99–110. Washington, DC: Georgetown University Press.

Lederer, S. E. 2004. "Military Personnel as Research Subjects." In *Encyclopedia of Bioethics*, edited by S. G. Post, 1843–45. New York, NY: Macmillan Reference USA.

Lee, P. 1981. "Permanence of the Ten Commandments: St. Thomas and His Modern Commentators." *Theological Studies* 42:422–43.

Lee, P. 1996. *Abortion and Unborn Human Life*. Washington, DC: The Catholic University of America Press.

Lee, P. 2001. "Personhood, Dignity, Suicide, and Euthanasia." *The National Catholic Bioethics Quarterly* 1(3):329–43.

Lee, P., and R. P. George. 2008. *Body-Self Dualism in Contemporary Ethics and Politics*. New York, NY: Cambridge University Press.

Leikin, S. 2003. "Minors' Assent, Consent, or Dissent to Medical Research." In *Ethical and Regulatory Aspects of Clinical Research: Readings and Commentary*, edited by E. J. Emanuel, R. A. Crouch, J. D. Arras, J. D. Moreno, and Ch. Grady, 252–58. Baltimore, MD: Johns Hopkins University Press.

Levine, R. J. 1982. "Research on Prisoners: Why Not?." *IRB: Ethics and Human Research* 4(5):6.

Levine, R. J. 1995. "Research in Emergency Situation. The Role of Deferred Consent." *The Journal of the American Medical Association* 273(16):1300–302.

Levine, R. J. 1996. "International Codes and Guidelines for Research Ethics: A Critical Appraisal." In *The Ethics of Research Involving Human Subjects: Facing the 21st Century*, edited by H. Y. Vanderpool, 235–59. Frederick, MD: University Pub. Group.

Levine, R. J. 1999. "The Need to Revise the Declaration of Helsinki." *The New England Journal of Medicine* 341(7):531–34.

Levine, R. J. 2003a. "Consent Issues in Human Research." In *Ethical and Regulatory Aspects of Clinical Research: Readings and Commentary*, edited by E. J. Emanuel, R. A. Crouch, J. D. Arras, J. D. Moreno, and Ch. Grady, 197–201. Baltimore, MD: Johns Hopkins University Press.

Levine, R. J. 2003b. "Research and Practice." In *Ethical and Regulatory Aspects of Clinical Research: Readings and Commentary*, edited by E. J. Emanuel, R. A. Crouch, J. D. Arras, J. D. Moreno, and Ch. Grady, 103–107. Baltimore, MD: Johns Hopkins University Press.

Levine, R. J. 2005. "The National Commission's Ethical Principles, with Special Attention to Beneficence." In *Belmont Revisited: Ethical Principles for Research with Human Subjects*,

edited by J. F. Childress, E. M. Meslin, and H. T. Shapiro, 126–35. Washington, DC: Georgetown University Press.
Liacos, P. J. 1989. "Is 'Substituted Judgment' a Valid Legal Concept?." *Issues in Law and Medicine* 5:215–24.
Lifton, R. J. 2000. *The Nazi Doctors: Medical Killing and the Psychology of Genocide*. New York, NY: Basic Books.
Lockhart, J. D. 1977. "Pediatric Drug Testing: Is It at Risk?." *Hastings Center Report* 7(3):8–10.
MacIntyre, A. C. 1999. *Dependent Rational Animals: Why Human Beings Need the Virtues*. Chicago, IL: Open Court.
Macklin, R. 2004. *Double Standards in Medical Research in Developing Countries*. Cambridge and New York, NY: Cambridge University Press.
Maher, D. 2001. "Parental Love and Prenatal Diagnosis." *The National Catholic Bioethics Quarterly* 1(4):519–26.
Mauron, A. 2004. "Embryo and Fetus. Development from Fertilization to Birth." In *Encyclopedia of Bioethics*, edited by S. G. Post, 707–12. New York, NY: Macmillan Reference USA.
May, W. E. 1974. "Experimenting on Human Subjects." *Linacre Quarterly* 41:238–52.
May, W. E. 1977. *Human Existence, Medicine, and Ethics: Reflections on Human Life*. Chicago, IL: Franciscan Herald Press.
May, W. E. 2000. *Catholic Bioethics and the Gift of Human Life* (1st ed.). Huntington, IN: Our Sunday Visitor Publishing Division.
May, W. F. 2001. "The Patient as Person: Beyond Ramsey's Beecher Lectures." In *The Patient as Person: Explorations in Medical Ethics*, edited by P. Ramsey, xxix–xliii. New Haven, CT: Yale University Press.
May, W. E. 2003a. *An Introduction to Moral Theology* (2nd ed.). Huntington, IN: Our Sunday Visitor Publishing Division.
May, W. E. 2003b. "Human Dignity and Biomedical Research: The Respective Positions of the Subject of Research and the Researcher." In *Ethics of Biomedical Research in a Christian Vision. Proceedings of the Ninth Assembly of the Pontifical Academy for Life (24–26 February, 2003)*, edited by J. De Dios Vial Correa and E. Sgreccia, 172–90. Vatican City: Libreria Editrice Vaticana.
May, W. E. 2007. "Proxy Consent for Nontherapeutic Experimentation." *The National Catholic Bioethics Quarterly* 7(2):239–47.
May, W. E. 2008. *Catholic Bioethics and the Gift of Human Life* (2nd ed.). Huntington, IN: Our Sunday Visitor Publishing Division.
McCormick, R. A. 1975. "Notes on Moral Theology." *Theological Studies* 36:77–129.
McCormick, R. A. 1979. "Reflections on the Literature." In *Readings in Moral Theology, No. 1. Moral Norms and Catholic Tradition*, edited by Ch. E. Curran and R. A. McCormick, 294–340. New York, NY: Paulist Press.
McCormick, R. A. 1981a. *Notes on Moral Theology: 1965 Through 1980*. Washington, DC: University Press of America.
McCormick, R. A. 1981b. "Proxy Consent in the Experimentation Situation." In *How Brave a New World? Dilemmas in Bioethics*, 51–71. Washington, DC: Georgetown University Press.
McCormick, R. A. 1981c. "Sharing in Sociality: Children and Experimentation." In *How Brave a New World? Dilemmas in Bioethics*, 87–98. Washington, DC: Georgetown University Press.
McCormick, R. A. 1991. "Who or What Is the Preembryo?." *Kennedy Institute of Ethics Journal* 1(1):1–15.
McCormick, R. A. 1998. "Some Early Reactions to *Veritatis Splendor*." In *Readings in Moral Theology, No.10. John Paul II and Moral Theology*, edited by Ch. E. Curran and R. A. McCormick, 5–34. New York, NY and Mahwah, NJ: Paulist Press.
McCormick, R. A. 2006. *The Critical Calling*. Washington, DC: Georgetown University Press.
McInerny, R. M. 1997. *Ethica Thomistica: The Moral Philosophy of Thomas Aquinas* (revised ed.). Washington, DC: Catholic University of America Press.

McKibbon, A., A. Eady, and S. Marks. 1999. *Evidence-Based Principles and Practice*. Hamilton, ON: B.C. Decker Inc.

"Medicine". 2008. *Encyclopaedia Britannica Ultimate Reference Suite Edition.*

Melina, L., and C. A. Anderson, eds. 2006. *The Way of Love*. San Francisco, CA: Ignatius Press.

Michels, R. 2003. "Are Research Ethics Bad for Our Mental Health?." In *Ethical and Regulatory Aspects of Clinical Research: Readings and Commentary*, edited by E. J. Emanuel, R. A. Crouch, J. D. Arras, J. D. Moreno, and Ch. Grady, 234–37. Baltimore, MD: Johns Hopkins University Press.

Military Tribunal 1. 1950. "Case1. *United States v. Karl Brandt et al.*" In *Trials of War Criminals Before the Nuremberg Tribunals under Control Council Law 10*. Washington, DC: U.S. Government Printing Office.

Miller, F. G. 2005. "William James, Faith, and the Placebo Effect (Critical Essay)." *Perspectives in Biology and Medicine* 48(2):273–82.

Miller, W. E., and D. S. Kreiner. 2008. "Student Perception of Coercion to Participate in Psychological Research." *North American Journal of Psychology* 10(1):53–64.

Mitscherlich, A., and F. Mielke. 1949. *Doctors of Infamy; The Story of the Nazi Medical Crimes*. New York, NY: Henry Schuman.

Montague, P. 1980. "Comparative and Non-Comparative Justice." *The Philosophical Quarterly* 30(119):131–40.

Montgomery, J. 2001. "Informed Consent and Clinical Research with Children." In *Informed Consent in Medical Research*, edited by L. Doyal and J. S. Tobias, 173–81. London: BMJ Books.

Moreno, J. D. 2003. "Convenient and Captive Populations." In *Ethical and Regulatory Aspects of Clinical Research: Readings and Commentary*, edited by E. J. Emanuel, R. A. Crouch, J. D. Arras, J. D. Moreno, and Ch. Grady, 258–62. Baltimore, MD: Johns Hopkins University Press.

Morowitz, H. J., and J. S. Trefil. 1992. *The Facts of Life: Science and the Abortion Controversy*. New York, NY: Oxford University Press.

Morris, M. C. 2005. "An Ethical Analysis of Exception from Informed Consent Regulations." *Academic Emergency Medicine* 12(11):1113–19.

Morris, M. C., and R. Nelson. 2007. "Randomized Controlled Trials as Minimal Risk: An Ethical Analysis." *Critical Care Medicine* 35(3):940–44.

Murdoch, I. 1970. *The Sovereignty of Good*. London: Routledge & K. Paul.

National Bioethics Advisory Commission. 2003. "Research Involving Persons with Mental Disorders That May Affect Decisionmaking Capacity." In *Ethical and Regulatory Aspects of Clinical Research: Readings and Commentary*, edited by E. J. Emanuel, R. A. Crouch, J. D. Arras, J. D. Moreno, and Ch. Grady, 229–33. Baltimore, MD: Johns Hopkins University Press.

National Commission for the Protection of Human Subjects of Biomedical and Behavioral Research. 1977a. *Research Involving Children: Report and Recommendations. DHEW Publication No. (OS) 77-0004*. Washington, DC: US Government Printing Office.

National Commission for the Protection of Human Subjects of Biomedical and Behavioral Research. 1977b. "Statement of Commissioners Height, King, Louisell, Ryan and Seldin." In *Research Involving Children: Report and Recommendations. DHEW Publication No. (OS) 77-0004*, 123–33. Washington, DC: US Government Printing Office.

National Commission for the Protection of Human Subjects of Biomedical and Behavioral Research. 1979. *The Belmont Report* [On-line] Accessed August 11, 2008. http://ohsr.od.nih.gov/guidelines/belmont.html

National Conference of Catholic Bishops. 2001. *Ethical and Religious Directives for Catholic Health Care Services* [On-line] Accessed August 1, 2008. http://www.usccb.org/bishops/directives.shtml

National Research Act. 1974. *Title II, Public Law*, 93–348.

Nelson-Marten, P., and B. A. Rich. 1999. "A Historical Perspective of Informed Consent in Clinical Practice and Research." *Seminars in Oncology Nursing* 15(2):81–88.

Nussbaum, R. L., R. R. McInnes, H. F. Willard, and M. W. Thompson. 2004. *Thompson & Thompson Genetics in Medicine* (6th ed.). Philadelphia, PA: Saunders.

"Nürnberg Trials". 2008. *Encyclopaedia Britannica Ultimate Reference Suite Edition*.

O'Donnell, T. J. 1996. *Medicine and Christian Morality* (3rd ed.). Staten Island, NY: Alba House.

O'Mathúna, D. P. 2002. "Genetic Technology, Enhancement, and Christian Values." *The National Catholic Bioethics Quarterly* 2(2):277–95.

O'Rourke, K. D. 2000. *A Primer for Health Care Ethics: Essays for a Pluralistic Society* (2nd ed.). Washington, DC: Georgetown University Press.

O'Rourke, K. D. 2004. "Stem Cell Research. Prospects and Problems." *The National Catholic Bioethics Quarterly* 4(2):289–99.

O'Rourke, K. D. 2006. "The Embryo as Person." *The National Catholic Bioethics Quarterly* 6(1):241–51.

Pacholczyk, T. "The Ten Great Myths in the Debate over Stem Cell Research" [On-line] Accessed August 4, 2008. http://www.ncbcenter.org/stemcell.asp

Paul VI. 1968. *Encyclical Letter on the Regulation of Birth "Humanae Vitae"* [On-line] Available at: http://www.vatican.va/holy_father/paul_vi/encyclicals/documents/hf_p-vi_enc_25071968_humanae-vitae_en.html.

Pearson, H. 2002. "Dual Identities." *Nature* 417:10–11.

Pellegrino, E. D. 1985. "Thomas Percival's Ethics: The Ethics Beneath the Etiquette." In *Medical Ethics; or a Code of Institutes and Precepts Adapted to the Professional Conduct of Physicians and Surgeons*, edited by T. Percival, 22–23. Birmingham, AL: Division of Gryphon Editions, Ltd.

Pellegrino, E. D. 1999. "Human Embryos and the Stem Cell Controversy." *The NaProEthics Forum* 4.6, 2–3.

Pellegrino, E. D. 2001a. "Beneficence, Scientific Autonomy, and Self-Interest: Ethical Dilemmas in Clinical Research." In *Physician and Philosopher. The Philosophical Foundation of Medicine: Essays by Dr. Edmund Pellegrino*, edited by R. J. Bulger and J. P. McGovern, 191–200. Charlottesville, VA: Carden Jennings Publishing

Pellegrino, E. D. 2001b. "The Goals and Ends of Medicine: How Are They to Be Defined?." In *Physician and Philosopher. The Philosophical Foundation of Medicine: Essays by Dr. Edmund Pellegrino*, edited by R. J. Bulger and J. P. McGovern, 56–65. Charlottesville, VA: Carden Jennings Publishing.

Pellegrino, E. D. 2001c. "Toward a Virtue-Based Normative Ethics for the Health Profession." In *Physician and Philosopher. The Philosophical Foundation of Medicine: Essays by Dr. Edmund Pellegrino*, edited by R. J. Bulger and J. P. McGovern, 113–29. Charlottesville, VA: Carden Jennings Publishing,

Pellegrino, E. D., and D. C. Thomasma. 1988. *For the Patient's Good: The Restoration of Beneficence in Health Care*. New York, NY: Oxford University Press

Pellegrino, E. D., and D. C. Thomasma. 1993. *The Virtues in Medical Practice*. New York, NY: Oxford University Press.

Pellegrino, E. D., and D. C. Thomasma 1996. *The Christian Virtues in Medical Practice*. Washington, DC: Georgetown University Press.

Pence, G. E. 2004, *Classic Cases in Medical Ethics: Accounts of Cases That Have Shaped Medical Ethics, with Philosophical, Legal, and Historical Backgrounds*. Boston, MA: McGraw-Hill.

Percival, T. 1985. *Medical Ethics; or a Code of Institutes and Precepts Adapted to the Professional Conduct of Physicians and Surgeons*. Birmingham, AL: Division of Gryphon Editions, Ltd.

Perley, S., S. S. Fluss, Z. Bankowski, and F. Simon. 1992. "The Nuremberg Code: An International Overview." In *The Nazi Doctors and the Nuremberg Code*, edited by G. J. Annas and M. A. Grodin, 149–73. New York, NY: Oxford University Press.

Pernick, M. S. 1982. "The Patient's Role in Medical Decisionmaking: A Social History of Informed Consent in Medical Therapy." In President's Commission for the Study of Ethical Problems in Medicine and Biomedical and Behavioral Research, *Making Health Care Decisions*, vol. 3. Washington, DC: U.S. Government Printing Office.

Peterson, E. 2004. "Key Legal Cases in Bioethics." In *Encyclopedia of Bioethics*, edited by S. G. Post, 2921–26. New York, NY: Macmillan Reference USA.

Pharmaceutical Research and Manufacturers of America. 2001. *PhRMA Discussion Paper on the Declaration of Helsinki as Revised in October 2000*. Washington, DC: Pharmaceutical Research and Manufacturers of America.

Piantadosi, S. 2005. *Clinical Trials: A Methodologic Perspective* (2nd ed.). Hoboken, NJ: Wiley.

Pieper, J. 1966. *The Four Cardinal Virtues: Prudence, Justice, Fortitude, Temperance*. Notre Dame, IN: University of Notre Dame Press.

Pinckaers, S. 1995. *The Sources of Christian Ethics*. Washington, DC: Catholic University of America Press.

Pinckaers, S. 2005a. "Aquinas on the Dignity of the Human Person." In *The Pinckaers Reader: Renewing Thomistic Moral Theology*, J. Berkman and C. S. Titus, 144–63. Washington, DC: Catholic University of America Press.

Pinckaers, S. 2005b. "Conscience and the Virtue of Prudence." In *The Pinckaers Reader: Renewing Thomistic Moral Theology*, edited by J. Berkman and C. S. Titus, 342–55. Washington, DC: Catholic University of America Press.

Pius XII. 1952. "To the Members of the First International Congress on Histopathology of the Nervous System." *Acta Apostolicae Sedis* 44:779–789.

Pius XII. 1957. "To the Doctors of the G. Mendel Institute." *Acta Apostolicae Sedis* 49:1027–1033.

"Placebo". 2000. In Paul Lagasse, *The Columbia Encyclopedia*. New York, NY: Columbia University Press.

"Placebo". 2002. In *Webster's Third International Dictionary, Unabridged* [On-line] Accessed August 8, 2008. http://unabridged.merriam-webster.com

Plich, R. 2006. "A Presentation and a Critique of T. L. Beauchamp and J. F. Childress's *Principles of Biomedical Ethics*." Unpublished doctoral dissertation. Washington, DC: The Pontifical John Paul II Institute for Studies on Marriage and Family at the Catholic University of America.

Pontifical Academy for Life. 2000. *Declaration on the Production of Embryonic Stem Cells for Scientific and Therapeutic Use* [On-line] Accessed September 4, 2008. http://www.vatican.va/roman_curia/pontifical_academies/acdlife/documents/rc_pa_acdlife_doc_20000824_cellule-staminali_en.html

Pontifical Council for Justice and Peace. 2004. *Compendium of the Social Doctrine of the Church*. Washington, DC: Libreria Editrice Vaticana & United States Conference of Catholic Bishops.

Pontifical Council for Pastoral Assistance to Health Care Workers. 1995. *Charter for Health Care Workers*. Boston, MA: Pauline Books and Media.

Pontifical Council for the Family. 1998. *Declaration on the Decrease of Fertility in the World* [On-line] Accessed August 13, 2008. http://www.vatican.va/roman_curia/pontifical_councils/family/documents/rc_pc_family_doc_29041998_fecondita_en.html

Portney, L. G., and M. P. Watkins. 2000. *Foundations of Clinical Research: Applications to Practice*. Upper Saddle River, NJ: Prentice-Hall, Inc.

Posever, T. A., and T. Chelmow. 2001. "Informed Consent for Research in Schizophrenia: An Alternative for Special Studies." *IRB: Ethics and Human Research* 23(1):10–15.

"Power of Attorney". 2002. *Webster's Third New International Dictionary, Unabridged* [On-line] Accessed August 8, 2008. http://unabridged.merriam-webster.com

Prentice, D. A. 2001. "Brave New World of Genetic Engineering." *The National Catholic Bioethics Quarterly* 1(4):529–39.

President's Commission for the Study of Ethical Problems in Medicine and Biomedical and Behavioral Research. 1982. *Making Health Care Decisions: The Ethical and Legal Implications of Informed Consent in the Patient-Practioner Relationship*. Washington, DC: U.S. Government Printing Office.

President's Council on Bioethics. 2004. *Monitoring Stem Cell Research*. Washington, DC.

Prieur, M. R. 1996. "Charter for Health Care Workers: A Fresh Face for Catholic Bioethics Teachings." *Catholic Health Association of Canada Review* 3(24):14–15.

Pritchard, I. A., and G. Koski. 2004. "Students as Research Subjects." In *Encyclopedia of Bioethics*, edited by S. G. Post, 2469–74. New York, NY: Macmillan Reference USA.
Rahner, K. 1972. "The Problem of Genetic Manipulation." In *Theological Investigations*, vol. 9, 225–52. New York, NY: Seabury Press.
Ramsey, P. 1961. *War and the Christian Conscience; How Shall Modern War Be Conducted Justly?* Durham, NC: Duke University Press.
Ramsey, P. 1962. *Nine Modern Moralists*. Englewood Cliffs, NJ: Prentice-Hall.
Ramsey, P. 1968. *The Just War: Force and Political Responsibility*. New York, NY: Scribner.
Ramsey, P. 1976. "The Enforcement of Morals: Nontherapeutic Research on Children." *The Hastings Center Report* 6:21–30.
Ramsey, P. 1978. *Ethics at the Edges of Life: Medical and Legal Intersections*. New Haven, CT: Yale University Press.
Ramsey, P. 1993. *Basic Christian Ethics*. Louisville, KY: Westminster/John Knox Press.
Ramsey, P. 2002. *The Patient as Person: Explorations in Medical Ethics* (2nd ed.). New Haven, CT: Yale University Press.
Ratzinger, J. 2007. "Conscience and Truth." In *On Conscience*, edited by J. Ratzinger, 11–41. San Francisco, CA: Ignatius Press.
Rawls, J. 1971. *A Theory of Justice*. Cambridge, MA: Harvard University Press.
Resnik, D. B. 2004. "Genetic Engineering, Human." In *Encyclopedia of Bioethics*, edited by S. G. Post, 959–66. New York, NY: Macmillan Reference USA.
Richardson, H. 2003. "Moral Reasoning." In *Stanford Encyclopedia of Philosophy* [On-line] Accessed August 14, 2008. http://plato.stanford.edu/entries/reasoning-moral
Ross, W. D. 1988. *The Right and the Good*. Indianapolis, IN: Hackett Pub. Co.
Rothman, K. J., and K. B. Michels. 1994. "The Continuing Unethical Use of Placebo Controls." *The New England Journal of Medicine* 331(6):394–98.
Rousselot, P. 2001. *The Problem of Love in the Middle Ages: A Historical Contribution*. Eng. trans. A. Vincelette. Milwaukee, WI: Marquette University Press.
Sass, H.-M. 1988. "Comparative Models and Goals for the Regulation of Human Research." In *The Use of Human Beings in Research: With Special Reference to Clinical Trials*, edited by S. F. Spicker, 47–85. Dordrecht and Boston, MA: Kluwer Academic Publishers.
Schaffner, K. F. 2004. "Research Methodology. Conceptual Issues." In *Encyclopedia of Bioethics*, edited by S. G. Post, 2326–47. New York, NY: Macmillan Reference USA.
Schindler, D. C. 2002. "Freedom Beyond Our Choosing: Augustine on the Will and Its Objects." *Communio* 29(4):618–53.
Serrao, D. 2003. "Ethics of Experimental Research in Humans: Principles and Guidelines." In *Ethics of Biomedical Research in a Christian Vision. Proceedings of the Ninth Assembly of the Pontifical Academy for Life (24–26 February, 2003)*, edited by J. De Dios Vial Correa and E. Sgreccia, 191–99. Vatican City: Libreria Editrice Vaticana.
Shamoo, A. E., and F. A. Khin-Maung-Gyi. 2002. *Ethics of the Use of Human Subjects in Research: Practical Guide*. New York, NY: Garland Pub.
Shapiro, H. T., and E. M. Meslin. 2005. "Relating to History: The Influence of the National Commission and Its *Belmont Report* on the National Bioethics Advisory Commission." In *Belmont Revisited: Ethical Principles for Research with Human Subjects*, edited by J. F. Childress, E. M. Meslin, and H. T. Shapiro, 55–73. Washington, DC: Georgetown University Press.
Shatzmiller, J. 1994. *Jews, Medicine, and Medieval Society*. Berkeley, CA: University of California Press.
Sherwin, S. 2005. "Belmont Revisited Through a Feminist Lens." In *Belmont Revisited: Ethical Principles for Research with Human Subjects*, edited by J. F. Childress, E. M. Meslin, and H. T. Shapiro, 148–64. Washington, DC: Georgetown University Press.
Sia, J., S. Paul, R. M. Martin, and H. Cross. 2004. "HIV Infection and Zidovudine Use in Childbearing Women." *Pediatrics* 114(6):707–12.

Simon, J. G., Ch. W. Powers, and J. P. Gunnemann. 1972. *The Ethical Investor*. New Haven, CT: Yale University Press.

Simon, M. 1845. *Déontologie Médicale Ou Des Devoirs et Des Droits Des Médecins Dans L'etat Actuel De La Civilisation*. Paris: J.B. Baillière.

Sims, M. 2005. "On-Line Physician-Patient Care." *Ethics & Medics* 30(2):1–3.

Sirotin, N., and B. Lo. 2006. "The End of Therapeutic Privilege?." *The Journal of Clinical Ethics* 17(4):312–16.

Smith, B. 2007. "A Kantian Analysis of Embryonic Stem Cell Research." *The National Catholic Bioethics Quarterly* 7(2):257–62.

Smith, M. A. 1995. *Human Dignity and the Common Good in the Aristotelian-Thomistic Tradition*. Lewiston, NY: The Edwin Mellen Press.

Sperling, R. S., D. E. Shapiro, R. W. Coombs, J. A. Todd, S. A. Herman, G. D. McSherry, M. J. O'Sullivan, R. B. Van Dyke, E. Jimenez, Ch. Rouzioux, P. M. Flynn, J. L. Sullivan, S. A. Spector, C. Diaz, J. Rooney, J. Balsley, R. D. Gelber, E. M. Connor, and AIDS Clinical Trials Group Protocol Study Group The Pediatric. 1996. "Maternal Viral Load, Zidovudine Treatment, and the Risk of Transmission of Human Immunodeficiency Virus Type 1 from Mother to Infant." *The New England Journal of Medicine* 335(22):1621–29.

Spitz, V. 2005. *Doctors from Hell: The Horrific Account of Nazi Experiments on Humans*. Boulder, CO: Sentient Publications.

Stringer, J. S. A., and D. J. Rouse. 1999. "Rapid Testing and Zidovudine Treatment to Prevent Vertical Transmission of Human Immunodeficiency Virus in Unregistered Parturients: A Cost-Effectiveness Analysis." *Obstet Gynecol* 94(1):34–40.

Sugarman, J. 2007. "Examining the Provisions for Research without Consent in the Emergency Setting." *The Hastings Center Report* 37(1):12–13.

Sunderland, T., and R. Dukoff. 1999. "Surrogate Decision Making and Advance Directives with Cognitively Impaired Research Subjects." In *Ethics in Psychiatric Research. A Resource Manual for Human Subjects Protection*, edited by H. A. Pincus, J. Pincus, J. A. Lieberman, and S. Ferris, 107–27. Washington, DC: American Psychiatric Association.

Surbled, G. 1905. *La Morale Dans Ses Rapports Avec La Médecine at L'hygiène*, vol. 3. Paris: V. Retaux et fils.

"Surrogate". 2002. *Webster's Third New International Dictionary, Unabridged* [On-line] Accessed August 8, 2008. http://unabridged.merriam-webster.com

Tauer, C. A. 2004. "Embryo and Fetus. Embryo Research." In *Encyclopedia of Bioethics*, edited by S. G. Post, 712–22. New York, NY: Macmillan Reference USA.

Taylor, T. 1992. "Opening Statement of the Prosecution. December 9, 1946." In *The Nazi Doctors and the Nuremberg Code: Human Rights in Human Experimentation*, edited by G. J. Annas and M. A. Grodin, 67–93. New York, NY: Oxford University Press.

"The Catholic Principles and Guidelines for Clinical Research". 2007. *The National Catholic Bioethics Quarterly* 7(1):152–65.

"The Nuremberg Code". 1998. In *Source Book in Bioethics*, edited by A. R. Jonsen, R. M. Veatch, and L. Walters, 11–12. Washington, DC: Georgetown University Press.

Thompson, Ch. 1989. "Affective Disorders." In *The Instruments of Psychiatric Research*, edited by Ch. Thompson, 87–126. New York, NY: Wiley

Tollman, S. M. 2001. "Fair Partnerships Support Ethical Research." *British Medical Journal* 323:1417–19.

Tonelli, M. R. 1997. "Substituted Judgment in Medical Practice: Evidentiary Standards on a Sliding Scale." *Journal of Law, Medicine & Ethics* 25(1):22–29.

Tonti-Filippini, N. 2007. "The Need for Ethics Committees, and Their Role and Function." *The National Catholic Bioethics Quarterly* 7(4):749–69.

Toulmin, S. 1976. "Fetal Experimentation: Moral Issues and Institutional Controls." In National Commission for the Protection of Human Subjects of Biomedical and Behavioral Research, *Appendix: Research on the Fetus. DHEW Publication No. (OS) 76-128*, sec. 10, 1–26. Washington, DC: US Government Printing Office.

Truog, R. D., W. Robinson, A. Randolph, and A. Morris. 1999. "Is Informed Consent Always Necessary for Randomized, Controlled Trials?." *The New England Journal of Medicine* 340(1):804–807.

Turner, B. J., C. J. Newschaffer, D. Zhang, T. Fanning, and W. W. Hauck. 1999. "Translating Clinical Trial Results into Practice: The Effect of an AIDS Clinical Trial on Prescribed Antiretroviral Therapy for HIV-Infected Pregnant Women." *Annals of Internal Medicine* 130(12):979–86.

Van den Heever, P. 2005. "Pleading the Defence of Therapeutic Privilege." *South African Medical Journal* 95(6):420–21.

Vanderpool, H. Y. 1996. "Introduction and Overview: Ethics, Historical Case Studies, and the Research Enterprise." In *The Ethics of Research Involving Human Subjects: Facing the 21st Century*, edited by H. Y. Vanderpool, 1–30. Frederick, MD: University Pub. Group.

Varmus, H., and D. Satcher. 1997. "Ethical Complexities of Conducting Research in Developing Countries." *The New England Journal of Medicine* 337(14):1003–1005.

Veatch, R. 1979. "Three Theories of Informed Consent: Philosophical Foundations and Policy Implications." In National Commission for the Protection of Human Subjects of Biomedical and Behavioral Research, *The Belmont Report: Ethical Principles and Guidelines for the Protection of Human Subjects of Research. DHEW Publication No. (OS) 78-0014*, Appendix, vol. 2, sec. 26, 1–66. Washington, DC: US Government Printing Office.

Veatch, R. 1981. *A Theory of Medical Ethics*. New York, NY: Basic Books.

Veatch, R. M. 1984. "Autonomy's Temporary Triumph." *The Hastings Center Report* 14(5):38–40.

Veatch, R. 1996a. "From Nuremberg Through the 1990s: The Priority of Autonomy." In *The Ethics of Research Involving Human Subjects: Facing the 21st Century*, edited by H. Y. Vanderpool, 45–58. Frederick, MD: University Pub. Group.

Veatch, R. 1996b. "Which Grounds for Overriding Autonomy Are Legitimate?." *The Hastings Center Report* 26(6):42–43.

Veatch, R. 2003. *The Basics of Bioethics* (2nd ed.). Upper Saddle River, NJ: Prentice Hall.

Veatch, R. 2005. "Ranking, Balancing, or Simultaneity. Resolving Conflicts Among the *Belmont* Principles." In *Belmont Revisited: Ethical Principles for Research with Human Subjects*, edited by J. F. Childress, E. M. Meslin, and H. T. Shapiro, 184–204. Washington, DC: Georgetown University Press.

Weich, S. 1998. "The Cohort Study." *International Review of Psychiatry* 10:284–90.

Weijer, Ch. 1998. "Research Methods and Policies." In *Encyclopedia of Applied Ethics*, vol. 3, edited by R. Chadwick, 853–60. San Diego, CA: Academic Press.

Weithorn, L. A., and D. G. Scherer. 1994. "Children's Involvement in Research Participation Decisions: Psychological Considerations." In *Children as Research Subjects*, edited by M. A. Grodin and L. H. Glantz, 133–79. New York, NY: Oxford University Press.

Wettstein, R. M. 2004. "Competence." In *Encyclopedia of Bioethics*, edited by S. G. Post, 488–94. New York, NY: Macmillan Reference USA.

Wildes, K. W. 2000. *Moral Acquaintances: Methodology in Bioethics*. Notre Dame, IN: University of Notre Dame Press.

Winslade, W. J., and T. L. Krause. 1998. "The Nuremberg Code Turns Fifty." In *Ethics Codes in Medicine: Foundations and Achievements of Codification Since 1947*, edited by U. Tröhler, S. Reiter-Theil, and E. Herych, 140–62. Aldershot and Brookfield, VT: Ashgate Pub.

Wojtyla, K. 1981. *Love and Responsibility*. Eng. trans. H. T. Willetts. San Francisco, CA: Ignatius Press.

World Medical Association. 1964. *Declaration of Helsinki. Ethical Principles for Medical Research Involving Human Subjects* [On-line] Accessed August 11, 2008. http://www.wma.net/e/policy/b3.htm

World Medical Association. 1992. "Declaration of Helsinki. Versions I through IV." In *The Nazi Doctors and the Nuremberg Code: Human Rights in Human Experimentation*, edited by G. J Annas and M. A. Grodin, 331–42. New York, NY: Oxford University Press.

World Medical Association. 1996. *Declaration of Helsinki. Version V* [On-line] Accessed August 11, 2008. http://www.hku.hk/facmed/images/document/04research/institution/Declaration_of_Helsinki_1996_version.pdf

Worsfold, V. 1974. "A Philosophical Justification of Children's Rights." *Harvard Educational Review* 44(1):142–57.

Index

N

O

P

R